Chitosan

Exploring Its Diverse Applications in Medicine, Agriculture, and Environmental Science

nova science publishers

www.novapublishers.com

Biochemistry Research Trends Series

Natural Pigments: Sources, Properties, and Health Benefits
Rashidi Othman, PhD,
Qurratu Aini Mat Ali, PhD,
Razanah Ramya, PhD,
Farah Ayuni Mohd Hatta, PhD,
Wan Syibrah Hanisah Wan Sulaiman, PhD,
and Nur Hanie Mohd Latiff, PhD
ISBN: 979-8-90134-055-4
eBook ISBN: 979-8-90134-119-3

Healing from the Depths: Marine Metabolites in Disease Management
Pranav Kumar Prabhakar, PhD
and Tarun Kumar Upadhyay, PhD
(Editors)
ISBN: 979-8-89530-832-5
eBook ISBN: 979-8-89530-983-4

Understanding Biotransformation
Mohammad asin Mohammad, PhD
(Editor)
ISBN: 979-8-89530-095-4
eBook ISBN: 979-8-89530-141-8

Enzymes and DNA Recombinant Technology
Shafat Ahmad Latoo, M.S, PhD
ISBN: 979-8-89113-735-6
eBook ISBN: 979-8-89113-864-3

Lipases and their Role in Health and Disease
Vasudeo P. Zambare, PhD,
and Mohd. Fadhil Md. Din, PhD
(Editors)
ISBN: 979-8-89113-628-1
eBook ISBN: 979-8-89113-674-8

More information about this series can be found at

Denis Hurst
Editor

Chitosan

Exploring Its Diverse Applications in Medicine, Agriculture, and Environmental Science

nova
science publishers
www.novapublishers.com

DOI: https://doi.org/10.52305/GMQT0542

Additional color graphics may be available in the e-book version of this book.

Library of Congress Cataloging-in-Publication Data

ISBN: 979-8-90134-079-0 (Hardcover)
ISBN: 979-8-90134-218-3 (eBook)

Published by Nova Science Publishers, Inc. † New York

Contents

Chapter 1

Natural Polysaccharide Chitosan: Structural Features, Functional Characteristics and Application Prospects

Li Huang[1], PhD
Wenyan Liu[1]
Hui Zhang[1], MA
Song Jin[2,*]
Xiang Nan[1]
and Yanchao Wu[3], PhD

[1]Shaanxi Key Laboratory of Catalysis, School of Chemical and Environmental Science, Shaanxi University of Technology, Hanzhong, P. R. China
[2]Faculty of Chemistry, Biology and Geography,
Kyrgyz National University named after Ishenaly Arabaev, Bishkek, Kyrgyzstan
[3]Weihai Key Laboratory of Active Factor of Marine Products,
School of Chemistry and Chemical Engineering, Harbin Institute of Technology, Harbin, P. R. China

Abstract

Chitosan is a natural linear polysaccharide derived from the chitin. Its unique chemical structure features alternating D-glucosamine and N-acetyl-D-glucosamine through β-(1, 4)-glycosidic bonds, forming a linear semi-crystalline framework. The free amino groups in the molecular chain endow chitosan with cationic properties and pH responsiveness, enabling it to protonate in acidic environments and

[*] Corresponding Author's Email: 13232887338@163.com

In: Chitosan
Editor: Denis Hurst
ISBN: 979-8-90134-079-0

forming positively charged networks. The acetylation degree and molecular weight distribution of chitosan are key factors in regulating physicochemical properties. Different sources of chitosan, such as those from crustaceans, fungi, and insects, exhibit diverse application potentials due to their structural differences. Chitosan possesses excellent biocompatibility, degradability, antibacterial, immunomodulatory, antioxidant and anti-inflammatory properties. Accordingly, it has been widely applied in biomedicine, agriculture, the food industry, environmental protection, and cosmetics. This chapter systematically analyzes the structures and properties of chitosan and explores its application potential in various fields. Practical and feasible solutions have been proposed for challenges such as the stability of raw material supply, the consistency of batch performance, and the safety of functional modification. The chapter is beneficial to the comprehensive promotion of the industrialization process of chitosan and its cross-disciplinary innovative application.

Keywords: chitosan, structural characterization, biological activities, application

1. Introduction

Chitosan is a natural linear polysaccharide derived from chitin through deacetylation. It is widely found in the shells of crustaceans, fungal cell walls, and insect exoskeletons. As one of the few alkaline polysaccharides found in nature, chitosan possesses a molecular structure in which D-glucosamine and N-acetyl-D-glucosamine are linked alternately via β-(1, 4)-glycosidic bonds, forming a unique linear semi-crystalline structure. (Abdullahi et al., 2025; Carbone et al., 2025). Each three sugar units on the molecular chain contain a free amino group, which enables chitosan to protonate under acidic conditions, forming a positively charged network. It interacts with the static electricity of negatively charged biomolecules or microbial cell membranes. This structural characteristic gives chitosan both biocompatibility and functional modulability, making it a bridge between materials science and life sciences (Alamrani, 2025). With the growing global demand for renewable biological resources, chitosan due to its wide availability, biodegradability, and low immunogenicity has shown strategic value in tissue engineering, drug delivery, agricultural protection, and environmental materials, becoming an important candidate to replace traditional petrochemical materials.

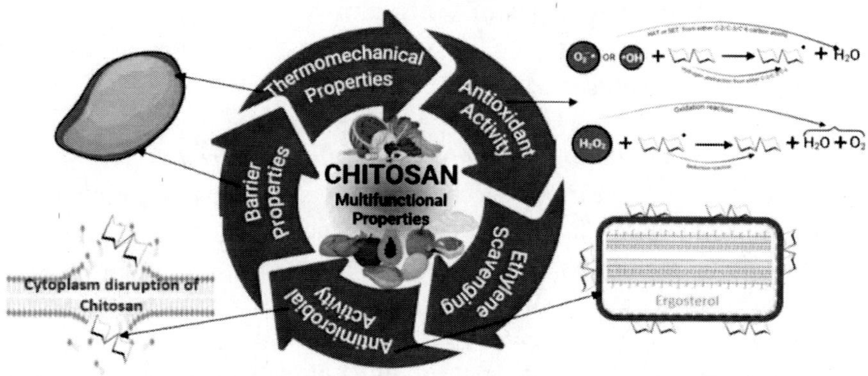

Figure 1. The biological activities of chitosan [Reproduced with permission from Elsevier, Ref. (Olunusi et al., 2025)].

Despite numerous advantages, chitosan faces multiple challenges in large-scale applications. The primary issues are the stability and quality control of raw materials. Crustacean chitosan relies on fishery waste and is subject to seasonal and regional restrictions, leading to price fluctuations and quality changes (Edo et al., 2025). Although fungal fermentation can enable industrial production, optimizing strains and fermentation processes to reduce costs remains a technical challenge. Additionally, the performance consistency of chitosan should be ensured through standardized characterization ways to address performance differences between batches (He et al., 2025). Another key challenge is the safety assessment of functional modifications. Because chemical modifications pose potential toxicity risks, strict toxicological tests are required to verify their biological safety. Considering these aspects, this chapter is intended to conduct a systematic analysis of chitosan's structure-property correlations, biological attributes, and a broad spectrum of its practical applications. The molecular chain conformation, functional group composition and crystal structures are revealed by advanced characterization technology, which provides a theoretical basis for functional modification. Moreover, its biological characteristics are described in detail. Besides, the chapter also explores the application potential of chitosan in biomedicine, agriculture, food, and environmental protection (Figure 1), analyzing the challenges and proposing solutions of chitosan. Through interdisciplinary studies, the chapter aims to facilitate the transition of chitosan from laboratory studies to industrial applications, providing scientific guidance and technical support for the development of high-performance and multifunctional chitosan-based materials.

2. Structure and Properties of Chitosan

2.1. Structure of Chitosan

2.1.1. Basic Molecular Composition and Chain Conformation

Chitosan is a natural linear polysaccharide derived from partial or complete deacetylation of chitin. Its molecular backbone is composed of *β*-1, 4-glycosidic bonds covalently linking two pyranose ring monomer units: D-glucosamine (GlcN) and N-acetyl-D-glucosamine (GlcNAc) (Iqbal et al., 2025; Jin et al., 2025). The C2 position of GlcN unit carries a free amino group, while the C2 position of GlcNAc unit carries an acetylamino group. The molar ratio of these two units is determined by the deacetylation degree (DDA), and they are typically arranged in random, block, or gradient sequences to form a heterogeneous copolymer chain. (Kumar et al., 2025). This chemical composition endows chitosan with amphiphilicity: The free amino and hydroxyl groups located at C3 and C6 positions provide hydrophilicity and cationic properties, while the acetyl groups enhance hydrophobic interactions. Regarding chain conformation, the pyranose ring adopts an energetically favorable chair conformation, while the β-glycosidic linkages give rise to a rigid helical architecture. Moreover, intramolecular hydrogen bonds between the C3 hydroxyl group and the O5 atom of the neighboring sugar ring serve to further stabilize this secondary structure. (Li et al., 2025b; Liu et al., 2025a). The bond length is approximately 2.8 Å. This structure results in low chain flexibility, with a persistence length of about 12 nanometers, but it imparts high tensile strength, with a theoretical modulus up to 40 gigapascals or more (Martingo et al., 2025a).

The distribution of monomer sequences has a decisive impact on function. High-density GlcN block regions, such as those containing more than five consecutive GlcN units, significantly enhance their electrostatic binding ability with negatively charged substances due to local positive charge enrichment, with a charge density that can exceed 5 mmol/g (Nabil et al., 2025; Oliveira Júnior et al., 2025; Panahimehr et al., 2025). Meanwhile, GlcNAc blocks mediate protein adsorption or drug encapsulation through hydrophobic interactions. The C2 amino group and C3, C6 hydroxyl groups constitute multifunctional reaction sites, with the pKa value of the amino group being approximately 6.5 (Remesh et al., 2025). Through targeted modifications such as carboxymethylation and with a degree of substitution ranging from 0.6 to 1.2, solubility can be regulated. Carboxymethyl chitosan has a solubility of up to 200 mg/mL at pH 7, breaking through the limitation of natural chitosan, which is only soluble in acids (Sayyad et al., 2025). Quaternization modification introduces trimethylammonium groups, which elevates the zeta potential from +35 mV to +60 mV and enhances the antibacterial efficiency by threefold. This molecular-level designability serves as the core foundation for its cross-disciplinary applications.

2.1.2. Core Parameters: Degree of Deacetylation and Molecular Weight

Degree of deacetylation (DD) is the most critical quality indicator of chitosan, defined as the percentage of GlcN units in the molecular chain relative to the total number of sugar units, typically ranging from 55% to 99% (Soni et al., 2025; Usta et al., 2025). When the deacetylation degree (DD) exceeds 85%, the high density of amino groups markedly enhances its cationic properties, including antibacterial activity and gene delivery capability. A DD value below 70% hinders amino protonation due to the steric hindrance effect of acetyl groups, leading to decreased acidic solubility. Accurate determination of DD requires a combination of titration, fourier transform infrared spectroscopy, or nuclear magnetic resonance spectroscopy (Vosoughi et al., 2025). Titration calculates free amino acids through acid-base neutralization; Fourier transform infrared spectroscopy utilizes the ratio of amide I/II peaks at 1655 cm^{-1} and 1590 cm^{-1} for estimation (Yildirim et al., 2025). Nnuclear magnetic resonance spectroscopy analyzes the area ratio of H2 signal peaks of GlcN to methyl signal peaks of GlcNAc. In industrial applications, high DD chitosan, products with a DD value greater than 90%, are predominantly used in the biomedical field (Zeng et al., 2025). However, low-deacetylation-degree (low-DD) chitosan variants are suitable for sustained-release formulations or membrane separation technologies, owing to their low cost and high hydrophobicity.

Molecular weight (MW) is another key parameter, ranging from 5 kDa for chitosan oligosaccharides to 1000 kDa for high molecular weight polymers (Zheng et al., 2025). Molecular weight is determined by viscometry, gel permeation chromatography, or light scattering methods. Viscometry is calculated based on the Mark-Houwink equation. Gel permeation chromatography requires the combination of a multi-angle laser light scattering detector. Light scattering method directly obtains absolute molecular weight (Zhou et al., 2025b). High molecular weight chitosan, samples with molecular weights greater than 100 kDa, form a high viscosity solution due to long-chain entanglement, making it suitable for injectable gels or wound dressings. Low molecular weight types, materials with molecular weights below 50 kDa, are ideal carriers for anticancer drug delivery systems due to their strong permeability and high cell uptake efficiency (Wujcicki et al., 2025). Molecular weight and degree of deacetylation have a synergistic effect. High deacetylation degree (DD) combined with high molecular weight (MW) enhances crystallinity and mechanical properties yet impairs solubility; in contrast, low MW coupled with high DD achieves a balance between solubility and biological activity—a property that is highly desirable for tissue engineering scaffold design.

2.1.3. Advanced Structure and Morphology: Crystallization and Charge Characteristics

The crystal structure of chitosan is regulated by both degree of deacetylation and molecular weight. There are two main crystal forms: the anhydrous form I, belonging to the orthorhombic system, with chains tightly stacked through strong hydrogen bonds; and the hydrated form II, belonging to the monoclinic system, with water molecules embedded in a weakened hydrogen bond network (Xing et al., 2025; Yang et al., 2025a; Yang et al., 2026). When the degree of deacetylation is within the range of 70% to 80%, the crystallinity is the highest, and the X-ray diffraction pattern exhibits sharp characteristic peaks at 10 degrees and 20 degrees in the 2θ angle, indicating optimal sequence regularity of GlcN and GlcNAc. When the degree of deacetylation is higher than 90% or lower than 60%, the increase in sequence disorder leads to a decrease in crystallinity (Zhang et al., 2025c). High crystallinity enhances the mechanical strength and degradation resistance of the material but compromises its solubility. For medical-grade chitosan, the deacetylation degree (DD) is typically required to be controlled within the range of 75%–85% to achieve a balanced performance profile.

The charge characteristics of chitosan stem from the protonation ability of its free amino groups, rendering it a natural cationic polyelectrolyte. With a pKa value of approximately 6.3 to 6.5, in an environment with a pH value less than 6.5, the amino groups protonate into positively charged ammonium ions, which bind to negatively charged substances through electrostatic interactions, such as bacterial cell membranes, DNA phosphate backbones, or heavy metal ions (Mawale et al., 2025; Mohammadalipour et al., 2025; Moreira et al., 2026). The charge density is directly determined by the degree of deacetylation: when the degree of deacetylation is 95%, the charge density reaches 4.6 mmol/g, significantly higher than that of 3.2 mmol/g when the degree of deacetylation is 70% (Prabhakar and Pavani, 2025). This characteristic endows chitosan with broad-spectrum antibacterial activity, achieving antibacterial effects by disrupting the membrane potential of microorganisms. This property also renders it an efficient gene transfection vector, which can neutralize the negative charge of DNA to form stable nanocomplexes. In the field of water treatment, this characteristic aids in the adsorption of pollutants such as hexavalent chromium ions or Congo red dye.

2.2. Physicochemical Properties of Chitosan

2.2.1. Solubility and Solution Behavior

The solubility of chitosan is strictly dependent on an acidic environment. The core mechanism involves the protonation of free amino groups to form ammonium ions when the pH value is below 6.5 (Sharma et al., 2025). The positive charge generated by protonation triggers electrostatic repulsion between molecular chains, simultaneously disrupting the interchain hydrogen bond network, enabling it to dissolve in dilute solutions such as acetic acid, hydrochloric acid, or citric acid. The degree of deacetylation is a decisive factor for solubility: chitosan with a degree of deacetylation above 85% exhibits a solubility exceeding 50 mg/mL in 1% acetic acid (Silva et al., 2025); whereas when the degree of deacetylation is below 70%, the solubility sharply decreases to less than 10 mg/mL due to the hindrance of acetyl groups to amino protonation (Wang et al., 2025a). Molecular weight affects processability through viscosity: A sample with a molecular weight of 100 kDa exhibits a viscosity of approximately 200 mPa·s in 1% acetic acid, while a sample with a molecular weight of 500 kDa can exceed 2000 mPa·s (Zhang et al., 2025e). During dissolution, it is necessary to control the pH value below the critical point of 6.3 to 6.5, otherwise gelation precipitation will occur.

Chemical modification can overcome the limitation of acidic dissolution: carboxymethylation introduces hydrophilic carboxyl groups, enabling the product to achieve a high solubility of up to 150 mg/mL in neutral pH (Chovancová et al., 2025); quaternization modification achieves full pH range solubility through permanent cationic groups. Optimization of the solvent system is also crucial: an acetic acid-urea mixed solvent (volume ratio 8:2) can increase the solubility of high-deacetylation and high-molecular-weight chitosan to 80 mg/mL (Cignolo et al., 2025). To avoid molecular chain degradation, the dissolution temperature should be strictly restricted to below 60°C. These physicochemical characteristics exert a direct influence on its efficacy in practical applications such as drug delivery systems and tissue engineering scaffolds.

2.2.2. Rheological Properties and Viscosity

The chitosan solution exhibits typical pseudoplastic fluid behavior, with its core characteristic being a significant decrease in apparent viscosity as the shear rate increases, a phenomenon known as the shear thinning effect. Taking a 2% concentration, 85% degree of deacetylation acetic acid solution as an example, the viscosity is approximately 500 mPa·s at a low shear rate of 1 s^{-1} (Ashiq et al., 2025; Attri et al., 2025). When the shear rate increases to 100 s^{-1}, the viscosity rapidly drops to 50 mPa·s (Chevrier et al., 2025). Such rheological behavior arises from the disentanglement and oriented alignment of polymer chains under shear stress, thereby conferring unique advantages for biomedical applications, including lower injection resistance and improved structural stability enabled by viscosity recovery in a quiescent state. (Fang et al., 2025). Dynamic rheological testing further reveals the regulation of behavior patterns by concentration: when the concentration is below 1%, the solution exhibits a viscous-dominant state, with the loss modulus higher than the storage modulus; when the concentration exceeds 2%, it transitions to an elastic gel state, with the storage modulus surpassing the loss modulus. This phase transition threshold is instructive for the mechanical design of tissue engineering scaffolds.

The viscosity of chitosan solution is synergistically regulated by multiple factors. An increase in concentration leads to an exponential growth in viscosity, for example, a rise from 1% to 2% can cause the viscosity to jump from 50 mPa·s to 400 mPa·s (Gao et al., 2025; Ghattas et al., 2025; Ghosh et al., 2025). An increase in molecular weight also markedly enhances viscosity, with a twofold increase in molecular weight leading to an approximate fivefold rise in viscosity. (Hsu et al., 2025). The influence of degree of deacetylation cannot be ignored either, with a 10% increase in degree of deacetylation leading to a 30% to 50% increase in viscosity (Ibrahim et al., 2025). Among environmental factors, a temperature increase from 20°C to 60°C leads to a 40% to 60% decrease in viscosity; the addition of electrolytes such as 0.1 mol/L NaCl compresses the double layer and weakens the electrostatic repulsion of molecular chains, resulting in a viscosity reduction of over 50% (Kiaee Nasab Abkenar and Dehnavi, 2025). These complex rheological properties make it a key regulatory target in the design of 3D printing ink formulations for biology, requiring precise balancing of extrudability and molding fidelity.

2.3. Film-Forming Property and Material Mechanical Properties

Chitosan can be fabricated into transparent and flexible films via solution casting, and its film-forming mechanism relies on hydrogen bond networks between molecular chains as well as electrostatic crosslinking of protonated amino groups. The mechanical properties of the films are deeply regulated by structural parameters: When the degree of deacetylation is 90% and the molecular weight reaches 200 kDa, the tensile strength of the film can reach 80 MPa, with an elongation at break of about 15% (Jogaiah et al., 2025; Kedves and Kónya, 2025). As the deacetylation degree falls to 70%, diminished crystallinity leads to a reduction in mechanical strength to 40 MPa, whereas the elongation of the material is enhanced to 25%. (Milbreta et al., 2025). The introduction of plasticizers can significantly improve the flexibility and toughness of the material. For example, adding 20% glycerol increases the elongation at break to 50%, while sacrificing the tensile strength to 30 MPa (Patel et al., 2025). This trade-off relationship between strength and ductility needs to be carefully balanced in the design of flexible dressings.

Barrier properties are one of the core advantages of chitosan films. Its oxygen barrier properties are particularly prominent, with an oxygen transmission rate of less than 5 $cm^3/(m^2 \cdot d \cdot atm)$, significantly better than the 100 $cm^3/(m^2 \cdot d \cdot atm)$ level of polyethylene films (Prabhakar and Pavani, 2025; Rusho et al., 2025). This characteristic can effectively inhibit the growth of aerobic bacteria in food active packaging. However, its water vapor barrier properties are weaker, with a moisture vapor transmission rate of about 200 $g/(m^2 \cdot d)$, which may accelerate food dehydration or excessive evaporation of wound exudate (Sundari et al., 2025). To overcome this drawback, composite modification strategies are commonly employed: for instance, after nanocomposite modification with montmorillonite, the moisture vapor transmission rate can be reduced to 50 $g/(m^2 \cdot d)$, while the oxygen transmission rate is maintained below 10 $cm^3/(m^2 \cdot d \cdot atm)$. (Yonpiam et al., 2025). These performance optimizations make it irreplaceable in antibacterial fresh-keeping packaging and wound moisture regulation dressings.

3. Biological Characteristics of Chitosan

3.1. Biocompatibility

Chitosan, as a natural polysaccharide compound, has biocompatibility due to its unique chemical structure and metabolic characteristics. Its molecule is composed of D-glucosamine (GlcN) and N-acetyl-D-glucosamine (GlcNAc) alternately connected by β-(1, 4)-glycosidic bonds, forming a linear semi crystalline structure (Zhou et al., 2025c; Zhou et al., 2025e). The abundant hydroxyl (-OH) and amino ($-NH_2$) groups in the molecular chain endow it with non-toxic and non-immunogenic properties. Studies have demonstrated that chitosan can be degraded in vivo by specific enzymes (e.g., lysozyme and chitosanase) into low-molecular-weight chitosan oligosaccharides and glucosamine; these degradation products are ultimately metabolized into carbon dioxide, water, and urea, and then excreted via the respiratory and urinary systems, thus ensuring the biological safety of chitosan (Liu et al., 2025b; Mahaninia and Yan, 2025). Its molecular chain has strong rigidity and mechanical properties that match natural tissues, such as high strength and modulus, making it suitable for scenarios such as surgical sutures and bone tissue engineering scaffolds that require mechanical loads. In addition, the anticoagulant properties of chitosan are demonstrated by promoting endothelial cell adhesion and proliferation, with good blood compatibility, and significantly improved anti thrombotic ability after phosphatidylation modification (Martingo et al., 2025b). Regarding tissue compatibility, co-culture experiments of chitosan with neural stem cells and vascular endothelial cells have shown that the cells maintain normal morphology and vigorous proliferation activity, without any inflammatory responses or carcinogenic tendencies—indicating its excellent tissue integration ability.

The biocompatibility of chitosan has been verified through rigorous biological evaluation tests, including cytotoxicity tests, sensitization tests, blood compatibility tests, etc., all of which have shown its high safety (Peng et al., 2025; Rajeswari et al., 2025). In terms of mechanical compatibility, chitosan can be matched with the mechanical requirements of different tissues (such as bones, skin, cartilage) by adjusting its molecular weight or introducing chemical modifications (such as cross-linking, grafting). For example, low molecular weight chitosan is used for soft tissue repair, while high molecular weight chitosan is used for high-strength bone scaffolds (Wang et al., 2025b; Xenou et al., 2025). Within the realm of biomedical applications, chitosan serves as a key raw material for the preparation of artificial skin, biological dressings, and hemostatic agents. Notably, its capacity to promote wound repair and suppress scar formation delivers significant therapeutic benefits for patients with large-area burns. As a drug carrier, chitosan can achieve drug sustained release and improve targeted therapeutic effects by controlling the degradation rate (Zanotti et al., 2025). The combination of its antibacterial activity and biocompatibility gives chitosan unique advantages in the field of infection prevention and control, such as the preparation of antibacterial dressings, dental restoration materials, and implant coatings (Zhou et al., 2025d). In addition, the immunomodulatory activity of chitosan is demonstrated by activating macrophages and lymphocytes, further expanding its potential application in tumor immunotherapy (Yilmaz and Yilmaz, 2026). Briefly, the chemical structure, biodegradability, mechanical properties, and biosafety of chitosan together constitute its excellent biocompatibility, providing a solid foundation for biomedical applications.

3.2. Degradability

Chitosan, as a natural polysaccharide polymer material, its biodegradability is a core characteristic that has attracted much attention in the fields of biomedical and environmental materials. The degradation of chitosan is primarily accomplished through three distinct pathways: Biodegradation, chemical degradation, and physical degradation. (Putro et al., 2025; Rezaie et al., 2025). Biodegradation is the main way of degrading chitosan, which relies on the action of enzymes such as chitosan and lysozyme. These enzymes can specifically recognize the β-(1, 4)-glycosidic bonds in chitosan molecules and break them into oligosaccharides or monosaccharides through hydrolysis reactions (Mu et al., 2025; Nandhakumar et al., 2025). For example, lysozyme indirectly acts on the glycosidic bonds of chitosan by disrupting the peptidoglycan backbone of bacterial cell walls, achieving mild and controllable degradation. Chemical degradation is mediated by acidic agents or oxidants such as hydrochloric acid and hydrogen peroxide; this process involves the cleavage of glycosidic linkages, thereby producing chitosan fragments with a broad range of molecular weights. (Mehrabi-Khozani et al., 2025). The hydrogen peroxide oxidation method has become the preferred method for industrial degradation due to its mild reaction conditions and narrow molecular weight distribution of the products. Physical degradation utilizes energy forms such as radiation and ultrasound to destroy the molecular structure of chitosan (Lin et al., 2025). For example, gamma rays generate hydroxyl radicals to trigger glycosidic bond cleavage, while the mechanical force of ultrasound can efficiently break the molecular chain (Mandal et al., 2025). The degradation products of chitosan are glucosamine and N-acetylglucosamine, which can be metabolized and absorbed by the human body, participate in the synthesis of glycosaminoglycans, promote cell proliferation and tissue repair.

The degradation rate of chitosan is influenced by various factors, among which deacetylation degree is one of the key parameters. The higher the degree of deacetylation, the more free amino groups in the molecule. In an acidic environment, protonation forms a positive charge, enhancing the interaction with enzymes or oxidants and accelerating degradation. As a typical example, medical-grade chitosan featuring a deacetylation degree (DD) of 85%–95% undergoes predominant degradation through enzymatic hydrolysis by lysozyme after being implanted *in vivo*. (Maury et al., 2025). Chitosan with lower deacetylation degree is more easily recognized and hydrolyzed by lysozyme due to its higher content of acetylated glucosamine residues, while highly deacetylated chitosan has a slower degradation rate and is suitable for scenarios that require long-term support, such as bone tissue engineering scaffolds (Agustiany et al., 2025). In addition, the reaction temperature and pH value also significantly affect the degradation efficiency. As a typical case, the hydrogen peroxide oxidation approach achieves maximum reaction kinetics under acidic conditions (pH 3–5). Elevating the temperature can accelerate the degradation process, yet precautions should be taken to prevent carbonization of the resultant products induced by temperature overrun.. The molecular weight and crystallinity of chitosan also affect degradation behavior. Chitosan with low molecular weight or amorphous structure is more easily penetrated by enzymes or chemical reagents, resulting in a faster degradation rate (Basso et al., 2025). These characteristics enable chitosan to obtain specific molecular weight products by regulating degradation conditions according to application requirements, such as oligochitosan, which has unique advantages in drug carriers, moisturizers, and other fields.

The biodegradability of chitosan enables it to exhibit excellent adaptability in both natural environments and living organisms. In the natural environment, chitosan can be gradually decomposed through microbial action or chemical hydrolysis (Bahamonde-Duarte et al., 2025). Enzymes secreted by microorganisms such as bacteria and fungi in soil can target the glycosidic bonds of chitosan, degrade it into small molecules such as glucosamine, and ultimately convert it into carbon dioxide and water, avoiding environmental pollution. For example, chitosan based agricultural films can be completely degraded in soil within 6 months, significantly reducing plastic residue issues (Çatal et al., 2025). In addition, chitosan is prone to chemical hydrolysis in acidic water, accelerating glycosidic bond cleavage after amino protonation. This characteristic makes it suitable for harmless treatment of heavy metal adsorption-chitosan that adsorbs heavy metals releases pollutants through acid hydrolysis and degrades itself into non-toxic products (Coer et al., 2025). Within biological systems, the in vivo degradation behavior of chitosan exhibits a strong correlation with its deacetylation degree. These degradation products are not only metabolizable and absorbable by the human body but also able to participate in glycosaminoglycan biosynthesis, thereby exerting a positive regulatory effect on cell proliferation and tissue regeneration (Ćirić et al., 2025). For example, chitosan based wound dressings continuously release antibacterial components during the degradation process, while maintaining a moist wound environment and accelerating healing. In addition, the degradability of chitosan makes it an ideal drug carrier material, achieving sustained release of drugs by regulating the degradation rate, improving therapeutic efficacy, and reducing side effects.

3.3. Antibacterial Activity

The antibacterial activity of chitosan originates from the electrostatic interaction between the protonated amino group ($-NH^{+}$) in its molecular structure and the microbial cell membrane (Figure 2). Under acidic conditions ($pH < 6.3$), the amino protonation of chitosan forms a positive charge, which can absorb negatively charged bacterial surfaces, disrupt cell membrane integrity, increase permeability, and lead to leakage of cell contents (Edo et al., 2025; Fusco et al., 2025). As observed via transmission electron microscopy, the cell walls of chitosan-treated bacteria displayed distinct pore structures, leading to the leakage of intracellular substances including nucleic acids and proteins. For example, under pH 5.5 conditions, chitosan with a deacetylation degree (DD) of 88.7% and a concentration of 1.0% exhibited an almost 100% inhibition rate against *Staphylococcus aureus* and *Escherichia coli*. (Gopalakrishnan et al., 2025). In addition, chitosan can also penetrate into bacterial cells, bind to DNA, and inhibit the synthesis of RNA and proteins. Low molecular weight chitosan (<50 kDa) has a shorter chain length, making it easier to penetrate the cytoplasm and directly interfere with genetic material replication; High molecular weight chitosan forms a barrier through physical adsorption, preventing nutrient intake.

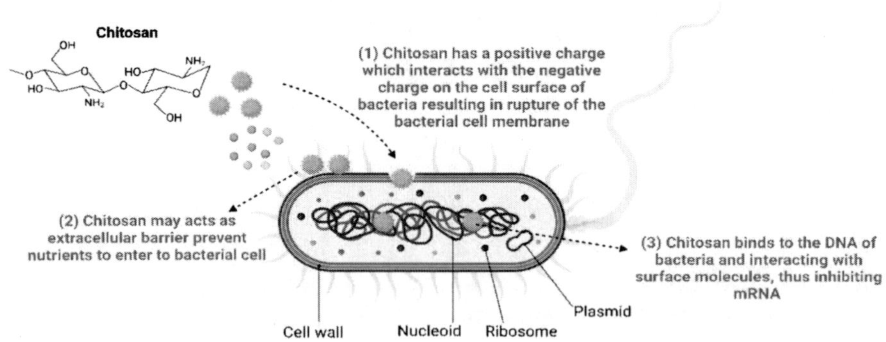

Figure 2. Antibacterial mechanisms of chitosan and chitosan derivatives [Reproduced with permission from Elsevier, Ref. (El-Saadony et al., 2025)].

The antibacterial activity of chitosan is regulated by multiple factors. The degree of deacetylation is a key parameter. The higher the degree of deacetylation, the more free amino groups there are, the stronger the protonation ability, and the more significant the antibacterial activity (Khodavandegar et al., 2025). For example, chitosan with a deacetylation degree of 81.56% has a much better antibacterial effect on Staphylococcus aureus than the sample with a deacetylation degree of 62.71%. Molecular weight also exerts an influence on the antibacterial mode of action: Low-molecular-weight chitosan exerts bactericidal effects via permeation, whereas high-molecular-weight chitosan functions through physical adsorption and membrane disruption. In terms of environmental conditions, acidic pH (<6.5) can enhance the solubility and positive charge density of chitosan, while increasing temperature can accelerate its reaction rate with cells (Kobisy et al., 2025). However, excessively high temperature ($>60°C$) may lead to molecular chain breakage and decreased activity. In addition, high concentrations of metal ions (Ca^{2+}, Mg^{2+}) competitively inhibit the antibacterial activity of chitosan, while anionic surfactants reduce activity by neutralizing positive charges.

In recent years, research on the antibacterial mechanism of chitosan has deepened to the level of gene expression and cellular signaling pathways. Research has found that chitosan can downregulate the expression of bacterial resistance related genes and reduce drug resistance (Ma et al., 2025b). For example, the recovery of antibiotic sensitivity in resistant strains treated with chitosan indicates that it can reverse the resistance mechanism. In addition, chitosan derivatives have significantly expanded their antibacterial spectrum through chemical modification. After introducing long-chain alkyl groups into quaternized chitosan, the density of quaternary ammonium groups increased to 2.5 mmol/g, and the minimum inhibitory concentration (MIC) against *Escherichia coli* decreased to 0.05 mg/mL (Mohamed et al., 2025). Carboxymethylation modification breaks through solubility limitations, increasing solubility by 85% under neutral conditions and increasing the inhibition rate of *Candida albicans* to 96% (Putra et al., 2025). In terms of metal complexes, when the size of silver nanoparticles in chitosan-silver complexes is controlled within the range of 15–25 nm, the antibacterial activity is enhanced threefold while cytotoxicity is reduced.

Frontier studies have also revealed the multi-target antibacterial properties of chitosan. Some derivatives can activate the reactive oxygen species (ROS) generation pathway, and flow cytometry detected that the ROS levels in the treatment group are 6.8 times higher than those in the control group, exacerbating bacterial oxidative stress (Sheikhlou et al., 2025). In addition, pH sensitive chitosan microspheres can rapidly release over 90% of the drug loading at the site of infection, achieving targeted antibacterial activity. The biofilm penetration enhancing derivative has a clearance rate of over 60% on mature biofilms, providing a new approach for the treatment of chronic infections (Serra et al., 2025). These advances not only deepen the understanding of the antibacterial mechanism of chitosan, but also provide theoretical support for the development of efficient and low toxicity antibacterial materials.

3.4. Immune Regulatory Activity

The immunomodulatory activity of chitosan is a highly concerned research direction in its biological activity, and its mechanism of action involves multi-level regulation of the innate and adaptive immune systems. Research has shown that chitosan can trigger immune responses by activating the complement system, enhancing the biological effects of complement proteins through classical or alternative pathways, thereby strengthening the body's non-specific defense against pathogens (Thathsarani et al., 2025; Vieira et al., 2025). For example, macrophages treated with chitosan exhibit significantly enhanced phagocytic activity, which is related to the stimulation of cell surface pattern recognition receptors by chitosan, thereby inducing the release of pro-inflammatory cytokines and nitric oxide (NO) (Wan et al., 2025). In addition, chitosan can directly act on lymphocytes, promote T cell proliferation and differentiation, enhance the killing activity of cytotoxic T lymphocytes (CTLs), and regulate the production of specific antibodies (Zhang et al., 2025d). Such bidirectional regulatory activity allows chitosan to bolster the defense mechanisms against immune deficiency while suppressing aberrant immune responses, including allergic reactions and autoimmune disorders.

The immunomodulatory activity of chitosan is significantly influenced by its physicochemical properties. Molecular weight is one of the key parameters. Low molecular weight chitosan (<50 kDa), due to its shorter chain length, is more likely to penetrate the cell membrane and enter the cytoplasm, directly binding to DNA to regulate gene expression, while high molecular weight chitosan activates immune cells through surface adsorption. The degree of deacetylation also affects activity (Nguyen et al., 2025; Osman et al., 2025). Chitosan with a deacetylation degree greater than 70% possesses a high density of free amino groups; thus, it can protonate under acidic conditions to carry a positive charge, thereby enhancing its interaction with negatively charged cell membranes. (Petchsoongsakul et al., 2025). Environmental conditions such as pH and temperature also significantly affect activity. Acidic environment can enhance the solubility of chitosan and its binding ability to cell receptors, while 37°C accelerates its signal transduction with immune cells (Saaid and Ibrahim, 2025). In addition, the immunomodulatory effect of chitosan also exhibits a dose-dependent relationship, with low doses (<100 μg/mL) mainly activating innate immunity, while high doses (>500 μg/mL) may induce immune tolerance.

In recent years, research on chitosan in the field of immune regulation has deepened to the level of molecular signaling pathways and nano delivery systems. By constructing chitosan-based nanocomposites (such as chitosan graphene oxide and chitosan carbon nanotubes), researchers have found that these carriers can significantly enhance the bioavailability of polysaccharide immunomodulators (Soussani et al., 2025). For example, the complex of chitosan graphene oxide loaded with Codonopsis pilosula polysaccharides can be targeted for delivery to macrophages, by activating the nuclear factor-kappa B (NF-κB) and mitogen-activated protein kinase (MAPK) signaling pathways, upregulating the expression of co stimulatory molecules, thereby enhancing antigen presentation ability (Smułka et al., 2025). Notably, chitosan derivatives such as quaternized chitosan and carboxymethyl chitosan exhibit an expanded immunomodulatory spectrum, which is achieved through the deliberate introduction of targeted functional groups during the modification process. Quaternary ammonium chitosan, due to its increased positive charge density, can more efficiently bind to pathogen associated molecular patterns, activate dendritic cells, and promote T helper 1 (Th1) type immune responses.

In terms of mechanism research, chitosan has been proven to intervene in the tumor immune microenvironment by regulating the expression of immune checkpoint molecules (Srinivasan et al., 2025). For example, low-molecular-weight chitosan can downregulate the expression of mannose receptors and arginase-1 on the surface of tumor-associated macrophages, inhibit their polarization toward the M2 phenotype, and upregulate programmed death-ligand 1 (PD-L1) expression to enhance the cytotoxic activity of T cells. In addition, the regulatory effect of chitosan on intestinal immunity has also received attention. It can improve the structure of intestinal microbiota, promote the production of short chain fatty acids, and activate the intestinal mucosal immune system (Zhang et al., 2025a). These findings not only deepen our understanding of the immune regulatory mechanism of chitosan but also provide theoretical support for its application in vaccine adjuvants, tumor immunotherapy, and inflammatory disease intervention.

3.5. Antioxidant

The antioxidant activity of chitosan is mainly reflected in its ability to scavenge free radicals and its inhibitory effect on oxidative stress (Figure 3). As a natural polysaccharide polymer material, chitosan can effectively neutralize various ROS, such as hydroxyl radical, superoxide anions (O_2^-), and 2, 2-diphenyl-1-picrylhydrazyl (DPPH) radicals, thereby slowing down the impact of oxidative damage on organisms or materials (Xu et al., 2025; Yang et al., 2025b). The antioxidant properties of chitosan are derived from the reducing hydroxyl (-OH) and amino ($-NH_2$) groups in its molecular structure; these functional groups can interrupt free radical chain reactions by donating electrons or hydrogen atoms, thereby alleviating oxidative damage. (Al-Amrah et al., 2025). For example, in the field of food preservation, chitosan coating can significantly inhibit lipid peroxidation and extend the shelf life of fruits and vegetables. In living organisms, chitosan can enhance the body's antioxidant defense ability by regulating the activity of antioxidant enzyme systems such as superoxide dismutase (SOD) and glutathione peroxidase (GSH Px).

The antioxidant properties of chitosan are regulated by various factors, among which molecular weight is one of the key parameters. Low molecular weight chitosan (<50 kDa) exhibits significantly enhanced antioxidant capacity due to the exposure of more active hydroxyl groups caused by main chain breakage (Bahndral et al., 2025). For example, the IC_{50} value of low molecular weight chitosan for DPPH radical scavenging is 0.63 mg/mL, while the scavenging rate of high molecular weight chitosan is less than 50% (Biswas et al., 2025). The degree of deacetylation also affects activity. Chitosan with a high degree of deacetylation (>80%) has a high density of free amino groups, which can protonate under acidic conditions and carry a positive charge, making it more effective in reacting with negatively charged free radicals. In terms of environmental conditions, acidic pH (<6.5) can enhance the solubility of chitosan and its contact efficiency with free radicals, while an increase in temperature (such as 37°C) can accelerate its reaction rate with free radicals (Chelminiak-Dudkiewicz et al., 2025). Furthermore, the addition of metal ions can synergistically boost the antioxidant performance of chitosan. A typical example is that the introduction of Cu^{2+} elevates the hydroxyl radical clearance rate of chitosan by 40%. The electrical properties of substituents also significantly affect activity (Choi et al., 2025). Introducing electron withdrawing groups (such as carboxymethyl) can enhance the electron cloud density of active groups and improve antioxidant properties, while electron withdrawing groups (such as acetyl) may reduce activity.

In recent years, research on chitosan in the field of antioxidant has expanded to the development of derivatives and nano delivery systems. Through chemical modification, researchers have synthesized various highly active chitosan derivatives, such as N-carboxyethyl chitosan and quaternized chitosan (Alzahrani et al., 2025; Castro et al., 2025). Quaternary ammonium chitosan, due to its increased positive charge density, can more efficiently bind to negatively charged free radicals while maintaining activity at neutral pH, expanding its application range. With respect to nano-delivery systems, chitosan-graphene oxide nanocomposites enable targeted delivery of antioxidants and improve their bioavailability. (Binder et al., 2025). For example, the chitosan graphene oxide complex loaded with Codonopsis pilosula polysaccharides enhances the antioxidant capacity of macrophages by activating the NF-κB signaling pathway. In addition, chitosan has also made progress in the field of biomedical applications, such as protecting neurons by regulating oxidative stress responses, or serving as a carrier for anti-tumor drugs to reduce oxidative damage caused by chemotherapy (Bernabé Vírseda et al., 2025). These studies not only deepen our understanding of the antioxidant mechanism of chitosan, but also provide theoretical support for its application in functional foods, cosmetics, and disease treatment.

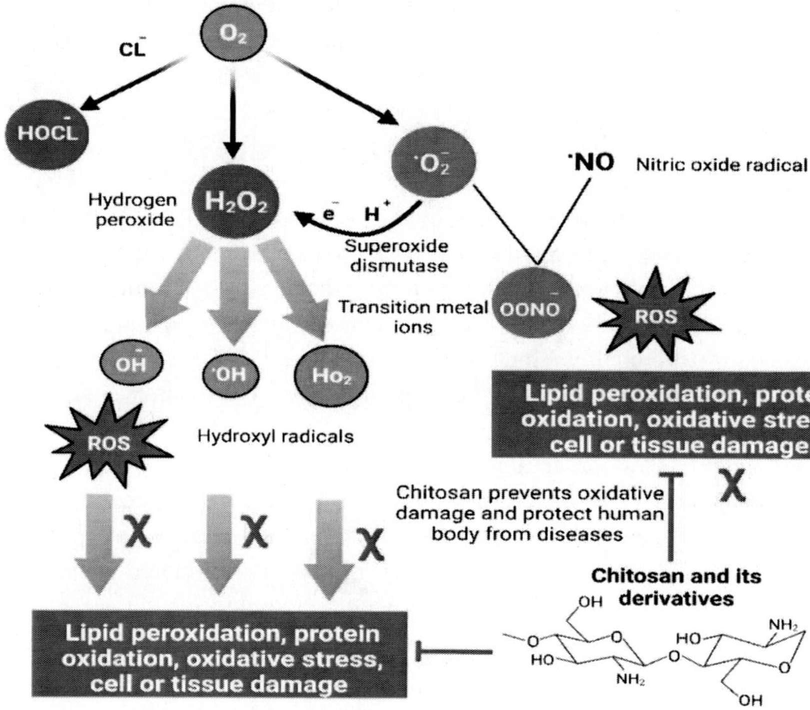

Figure 3. Antioxidant activity of chitosan and chitosan derivatives.
ROS, reactive oxygen species [Reproduced with permission from Elsevier, Ref. (El-Saadony et al., 2025)].

3.6. Anti-Inflammatory

The anti-inflammatory activity of chitosan is a highly concerned research direction in its biological activity, and its mechanism of action involves the regulation of inflammatory mediators, activation of immune cells, and inhibition of oxidative stress (da Silva Ponciano et al., 2025; Dejene and Abtew, 2025). Studies have demonstrated that chitosan can effectively inhibit the secretion of pro-inflammatory cytokines while upregulating the expression of anti-inflammatory factors (IL-10), thereby alleviating inflammatory responses. Its anti-inflammatory effect may be achieved through the following pathways: Firstly, by binding positively charged amino groups in the molecular structure to negatively charged bacterial cell walls, disrupting cell structure, inhibiting pathogen reproduction, and reducing inflammatory triggering factors (Demircivi et al., 2025). The second is to regulate the function of immune cells, such as activating macrophages and lymphocytes, enhancing their phagocytic and cytotoxic abilities, while suppressing excessive immune reactions and avoiding tissue damage (Deyab et al., 2025). For example, in arthritis models, chitosan significantly alleviates joint swelling and pain by reducing the concentration of inflammatory factors in joint fluid.

The anti-inflammatory properties of chitosan are regulated by multiple factors. Deacetylation degree is one of the key parameters. Chitosan with a high degree of deacetylation (>80%) has a high density of free amino groups, and after protonation under acidic conditions, it carries a positive charge, which can more effectively bind to negatively charged cell membranes or inflammatory mediators, thereby enhancing anti-inflammatory activity. Molecular weight also affects its mode of action (Esteve-Redondo et al., 2025; Fan et al., 2025). Low-molecular-weight chitosan (molecular weight < 50 kDa), owing to its shortened chain length, can more readily penetrate cell membranes and translocate into the cytoplasm, where it directly binds to DNA to regulate gene expression and suppress inflammation-related signaling pathways. (Gedik et al., 2025); High molecular weight chitosan activates immune cells through surface adsorption, indirectly exerting anti-inflammatory effects.

In recent years, research on chitosan in the field of anti-inflammatory has expanded to the development of derivatives and nano delivery systems. Through chemical modification, researchers have synthesized various highly active chitosan derivatives, such as lactic acid chitosan and quaternized chitosan (Ghorai et al., 2025). For example, lactate chitosan (CTL) improves metabolic associated fatty liver disease by activating the takeda G protein-coupled receptor 5-protein kinase A (TGR5-PKA) inflammatory suppression pathway, while regulating gut microbiota and bile acid metabolism, providing a new strategy for natural product therapy of metabolic diseases (Ghysels et al., 2025). In terms of nano delivery systems, nanocomposites can target the delivery of anti-inflammatory drugs, improve bioavailability, and enhance anti-inflammatory effects. For example, the chitosan graphene oxide complex loaded with Codonopsis pilosula polysaccharides enhances the antioxidant capacity of macrophages by activating the NF-κB signaling pathway (He and Wu, 2025). Furthermore, progress has been achieved in the utilization of chitosan within oral mucosal drug delivery platforms. By virtue of its mucoadhesion and permeability-promoting capacity, chitosan can extend the residence time of drugs, increase local drug levels, and ultimately enhance the therapeutic anti-inflammatory effects. These studies not only deepen our understanding of the anti-inflammatory mechanism of chitosan, but also provide theoretical support for its application in functional foods, cosmetics, and disease treatment.

4. Uses of Chitosan

4.1. The Biomedical Field

Chitosan, as a natural polysaccharide polymer material, has deeply penetrated into multiple aspects of clinical practice in the field of biomedical applications. One of its core applications is as a new type of wound dressing (Figure 4), which significantly improves the efficiency of wound repair by virtue of its hemostatic, antibacterial, and tissue regeneration promoting properties (Heidari et al., 2025). Traditional wound dressings merely provide physical coverage of wounds, whereas dressings fabricated from chitosan absorb exudate via their three-dimensional porous structure to maintain a moist wound environment, exhibiting superior performance compared with conventional gauze dressings for example, chitosan wound dressings can be used as adjuvant therapy for skin injuries such as shallow second degree burns and acne (Hussain et al., 2025). Their good biocompatibility and degradability, as well as the formation of a protective film, can reduce the risk of bacterial infection, stimulate collagen synthesis, and shorten the healing cycle. In addition, the antibacterial properties of chitosan can inhibit the growth of pathogenic bacteria such as Propionibacterium acnes, which is used in acne reducing skincare products.

In the field of drug delivery systems, chitosan also exhibits excellent drug loading performance (Figure 5). The amino and hydroxyl groups on its molecular chain provide multiple reaction sites, which can be used to prepare dosage forms such as microspheres and nanoparticles, achieving sustained release and targeted delivery of drugs (Iftime et al., 2025). Microspheres fabricated from chitosan exhibit favorable adhesive properties, making them suitable for mucosal administration routes such as oral, nasal, and gastrointestinal delivery, while improving the bioavailability of loaded drugs. For example, chitosan nanoparticles loaded with antibiotics adsorb onto bacterial surfaces through electrostatic interactions, enhancing the local concentration of the drug at the site of infection (Im et al., 2025). In addition, the mucosal adhesion of chitosan makes it widely used in ophthalmic preparations, nasal administration, and other fields, improving bioavailability by prolonging drug retention time. Chitosan can also be processed into films, hydrogels, fibers and other forms for the delivery of proteins/peptides, growth factors, anti-inflammatory drugs, anti-cancer drugs, vaccines, etc., as well as gene therapy (Figure 6). Currently, research on chitosan in the biomedical field is advancing towards cutting-edge directions such as tissue engineering and regenerative medicine, as well as the development of intelligent materials (Khainskaya et al., 2025). Within the realm of orthopedic regeneration, chitosan scaffolds offer an optimal milieu for cellular proliferation and differentiation, and can carry growth factors to promote the differentiation of bone-forming cells. At the same time, their anti-inflammatory properties weaken immune rejection reactions and promote bone tissue healing (Khani et al., 2025). For example, chitosan combined with natural bone tissue components such as hydroxyapatite can improve the mechanical properties of scaffold materials and meet the needs of bone tissue repair. In addition, chitosan has been widely used in cartilage tissue engineering, skin tissue engineering, and myocardial tissue engineering, providing a growth structural framework for damaged tissues and promoting tissue regeneration.

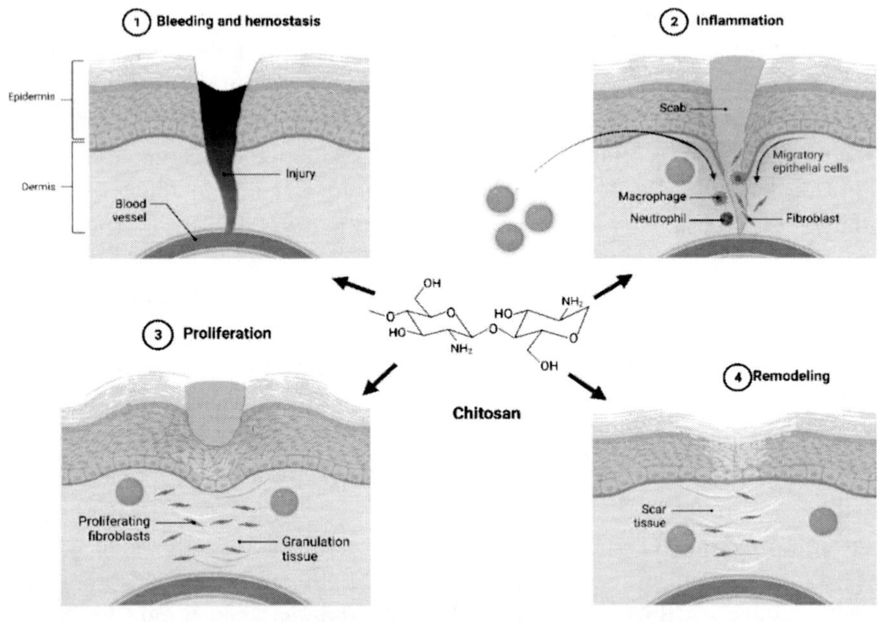

Figure 4. Chitosan and its role in facilitating the healing of wounded tissues [Reproduced with permission from Elsevier, Ref. (El-Saadony et al., 2025)].

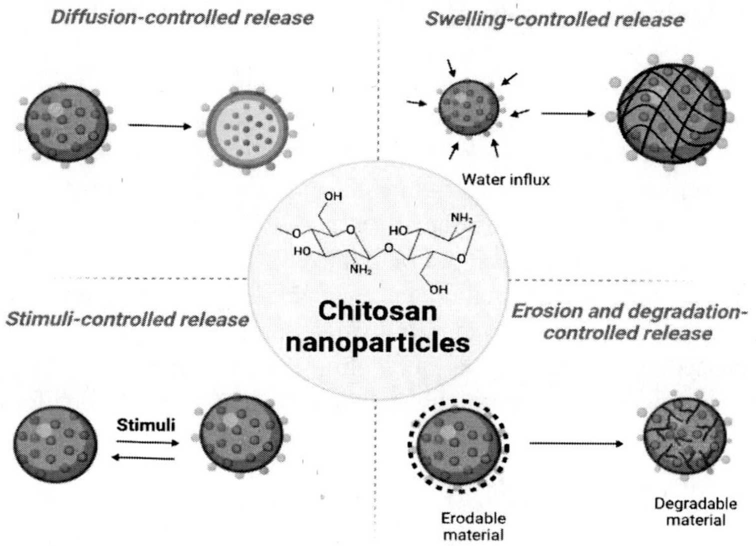

Figure 5. Methods for drug release from chitosan nanoparticles [Reproduced with permission from Elsevier, Ref. (El-Saadony et al., 2025)].

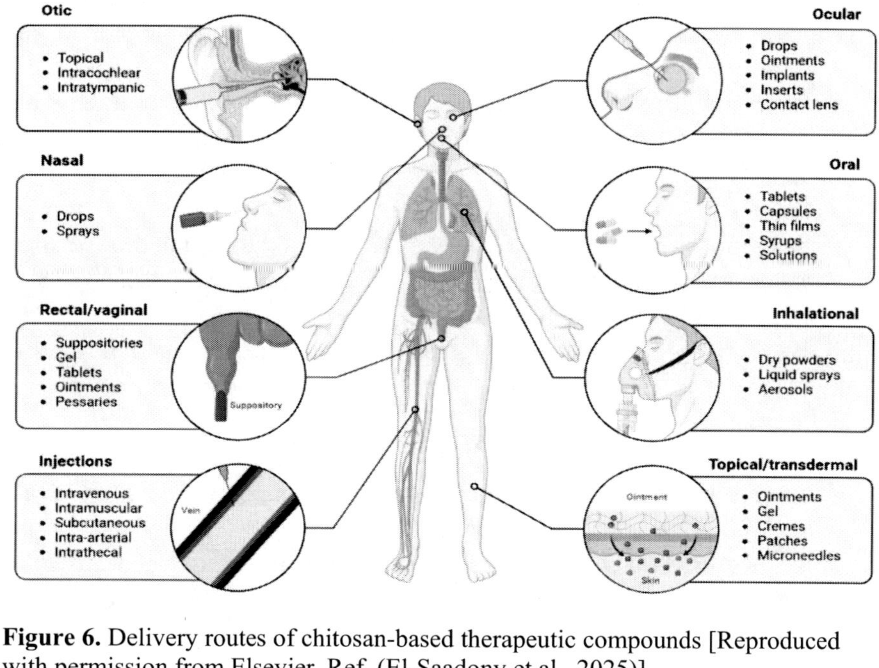

Figure 6. Delivery routes of chitosan-based therapeutic compounds [Reproduced with permission from Elsevier, Ref. (El-Saadony et al., 2025)].

4.2. The Agricultural Field

Chitosan, as a natural biopolymer, has shown significant value in the basic applications of agriculture, especially in plant growth regulation, stress resistance enhancement, and pest control. As a plant growth regulator, chitosan can promote seed germination and root development through soaking, coating, or foliar spraying (Li et al., 2025a). For example, soaking corn seeds in a 0.1% concentration chitosan solution for 12-24 h can significantly improve germination rate; After being coated with chitosan, soybean seeds showed improved germination potential and protein synthesis ability in seedling leaves. Furthermore, chitosan is capable of activating soil nutrient pools and forming protective coatings to achieve controlled fertilizer release—this not only extends the duration of fertilizer efficacy but also improves the uptake efficiency of nitrogen, phosphorus, potassium, and other essential elements by crops. (Li et al., 2025c). In terms of stress resistance, chitosan can enhance crops' resistance to adversity such as low temperature and drought. Its mechanism of action includes increasing soluble sugar content, enhancing photosynthesis, and forming a protective film to reduce water transpiration. For example, under cold conditions, wheat plants treated with chitosan have increased soluble sugar content and significantly enhanced cold resistance (Manu et al., 2025). With regard to pest management, chitosan has earned the title of "plant vaccine" for its ability to elicit the production of broad-spectrum defensive compounds in plants, while suppressing the spore germination and vegetative growth of pathogenic bacteria. For example, after spraying chitosan solution on tobacco leaves, the transmission inhibition rate of tobacco stripe virus is over 90% (Ma et al., 2025a). At the same time, it also has a repellent effect on pests such as nematodes and aphids, which can reduce the use of chemical pesticides. In addition, chitosan, as a soil amendment, can inhibit the proliferation of pathogenic bacteria, promote the proliferation of beneficial microorganisms, thereby improving soil aggregate structure, enhancing water retention and fertility.

At present, studies on chitosan in the agricultural field are centered on state-of-the-art research directions such as the remediation of soil heavy metal contamination and the development of smart materials, which underscore its innovative prospects for eco-friendly agricultural practices (Nath et al., 2025). In the field of heavy metal pollution remediation, chitosan has become a research hotspot due to its strong metal ion adsorption ability. For example, remediation agents fabricated from agricultural chitin via processes including microdissolution and modification can effectively immobilize heavy metal ions in soil, thereby reducing heavy metal uptake by crops. (Naveedunissa et al., 2025). Experimental data demonstrate that the application of chitosan remediation agents reduces the bioavailable cadmium content in soil by 62%, with heavy metal levels in crop grains subsequently meeting national food safety standards. In addition, the combination of chitosan and nanotechnology has promoted the research and development of intelligent slow-release fertilizers (Piccioni et al., 2025). For example, chitosan/urea nanoparticles achieve slow nutrient release through hydrogen bonding, increasing nitrogen utilization efficiency from 35% to 58% and significantly reducing water pollution caused by fertilizer loss. From the perspective of sustainable agriculture, notable progress has been made in developing biodegradable mulch films. These films, which take chitosan as the core raw material, can undergo biodegradation in soil environments, effectively eliminating the residual contamination associated with traditional plastic films. For example, in the corn planting experiment in Gansu, chitosan-based film increased soil temperature by 2-3°C, water retention rate by 40%, and achieved zero residue of the film (Van Overschelde et al., 2025). These innovative applications not only improve the efficiency of agricultural resource utilization but also provide new solutions for global food security and ecological protection.

4.3. The Food Industry

Chitosan, as a natural polysaccharide polymer material, has shown extensive application value in the food industry. Its basic applications include natural food antibacterial agents, fruit preservatives, food antioxidants, health food additives, juice clarifying agents, food thickening stabilizers, low-fat food fat substitutes, beverage clarifying and deacidification agents, and water purifiers (da Silva Bruckmann et al., 2025). As a natural food antibacterial agent, chitosan's positive charge interacts with the negative charge on bacterial cell membranes, disrupting cell structure and significantly inhibiting common foodborne pathogens such as *Escherichia coli* and *Staphylococcus aureus*, effectively extending the shelf life of food. With regard to fruit preservation, films fabricated from chitosan exert a multi-faceted effect: They modulate the respiratory metabolism of fruits and vegetables, suppress ethylene release, retard the ripening process, and form a physical protective layer that blocks microbial infection, thus maintaining the sensory freshness and nutritional integrity of the treated produce. (Duymaz et al., 2025). In addition, chitosan can chelate with free iron ions in meat, inhibit its catalytic activity, reduce food oxidation and spoilage, and maintain food color and flavor. As a health food additive, chitosan is considered a potential source of dietary fiber, which can bind with lipid compounds such as fat and cholesterol, reduce their absorption, and have health functions such as lipid-lowering and blood sugar lowering (Afsharikhah et al., 2025). In juice processing, chitosan serves as a clarifying agent, and its positive charge adsorbs and coagulates with negatively charged substances such as pectin and cellulose in the juice, achieving efficient clarification without affecting the nutritional content and taste of the juice (Apăvăloaiei et al., 2025). Meanwhile, chitosan can also serve as a food thickening stabilizer, improving the texture and taste of foods such as ice cream, mayonnaise, and bread, and extending their shelf life. In the development of low-fat foods, chitosan can bind with fat molecules to form a complex that is not easily absorbed by the human body, reducing fat intake and meeting consumers' demand for healthy foods (Athanasopoulou et al., 2025). In addition, chitosan can be applied to the clarification and deacidification of beverages including alcoholic drinks and fruit juices, where it removes impurities and turbidity, regulates beverage acidity, and enhances organoleptic properties. Finally, chitosan also performs well in water purification, more effectively removing pollutants such as polychlorinated biphenyls than activated carbon (Aman et al., 2025). When combined with bentonite to treat drinking water, it can remove particulate matter, color, and odor, and improve water quality.

With the continuous development of the food industry, the application fields of chitosan are also constantly expanding, and its cutting-edge exploration mainly focuses on functional food development, intelligent packaging materials, and food detection sensors (Crema et al., 2025). Within the realm of functional food innovation, chitosan has been extensively applied in the development of nutraceuticals, leveraging its distinctive physiological functions and processability to deliver products with hypolipidemic, antihyperglycemic, and immunomodulatory effects. For example, chitosan, as a dietary fiber and prebiotic, can promote intestinal health, improve gut microbiota structure, and enhance human immunity (Cibotaru et al., 2025). In addition, chitosan can also be combined with other active ingredients to develop composite functional foods with multiple health benefits. In terms of intelligent packaging materials, chitosan based intelligent packaging materials can respond to environmental changes (such as temperature and humidity) to release preservatives, extend food shelf life, and improve food safety (Davydova et al., 2025). This intelligent packaging material not only has the protective function of traditional packaging materials but also can automatically adjust the gas composition and humidity inside the packaging according to changes in the food storage environment, providing more suitable storage conditions for food. With regard to food sensing technologies, the combination of chitosan and nanotechnology has led to the fabrication of advanced food detection sensors, which enable the rapid and precise detection of hazardous compounds and microbial contaminants in food matrices, improving the overall efficacy of food safety surveillance. (Gunasekaran et al., 2025). This type of sensor has the advantages of high sensitivity, good selectivity, and fast response speed, providing a new technological means for food safety supervision. With the continuous development and improvement of these cutting-edge technologies, the application prospects of chitosan in the food industry will be even broader.

4.4. The Environmental Protection Field

Chitosan, as a natural polymer material, has shown extensive application value in the field of environmental protection. Its basic applications include wastewater treatment, heavy metal adsorption, soil remediation, and many other aspects (Hevira et al., 2025). Owing to its distinctive molecular structure, chitosan has emerged as a highly efficient flocculant for wastewater treatment. The amino groups on its molecular chain protonate under acidic conditions, forming positively charged polymers that can neutralize the negative charge carried by colloidal particles in wastewater, thereby disrupting the stability of colloids and causing them to aggregate and settle (Charles et al., 2025). In addition, the adsorption bridging effect of chitosan is also crucial. Its polymer chains can simultaneously adsorb multiple colloidal particles, forming a network structure of flocs, significantly improving the efficiency of wastewater treatment. In terms of heavy metal adsorption, the amino and hydroxyl groups in chitosan molecules can form stable complexes with heavy metal ions, effectively removing heavy metal ions from wastewater (Afrose et al., 2025). For instance, chitosan exhibits remarkable adsorption capacity for heavy metal ions (e.g., Cu^{2+} and Pb^{2+}), with the adsorption capacity exceeding 100 mg/g. In addition, chitosan can also act as a soil remediation agent, which optimizes soil structure and enhances soil fertility to create a favorable growth environment for plants.

With the continuous development of environmental protection technology, the application of chitosan in environmental protection is also constantly expanding and deepening (Gibeaux et al., 2025; Zhang et al.,

2025f). Currently, the cutting-edge exploration of chitosan mainly focuses on functional modification, intelligent material development, and environmentally friendly adsorption materials. The properties of chitosan have been significantly improved through chemical modification methods such as cross-linking and graft polymerization (Zhu et al., 2025). For example, after the composite of chitosan and graphene oxide, the adsorption capacity and stability of heavy metal ions are significantly improved. In addition, chitosan based smart materials have shown great potential in environmental monitoring and remediation. This category of materials can respond to environmental stimuli including pH and temperature fluctuations, thereby releasing adsorbents or catalysts to enable the dynamic capture of heavy metals and the degradation of organic pollutants. Meanwhile, chitosan has also made significant progress in the development of environmentally friendly adsorbent materials (Zhuravleva et al., 2026). For example, biodegradable membrane materials prepared based on chitosan not only have excellent heavy metal adsorption performance, but also do not produce secondary pollution during use, providing a new solution for heavy metal purification in water bodies.

4.5. The Cosmetics Field

As a natural polymeric material, chitosan exhibits distinctive application value in the cosmetics field, with its primary applications spanning multiple functional aspects including moisturization, skin repair, antibacterial activity, and hair care. In terms of moisturizing and repairing, chitosan can form a hydrophilic film on the surface of the skin, effectively locking in moisture, reducing water evaporation, and thus maintaining the skin's moist state (Zhang et al., 2025b). This moisturizing effect is similar to hyaluronic acid, but the moisturizing effect of chitosan is more long-lasting. In addition, chitosan also has the function of promoting wound healing and repairing the epidermal barrier, which can accelerate the recovery process of postoperative or damaged skin. In terms of antibacterial and anti-inflammatory properties, chitosan has natural antibiotic properties that can inhibit the growth of harmful microorganisms such as *Propionibacterium acnes* and alleviate skin inflammation (Ye et al., 2025). This has a significant effect on treating skin problems such as acne and dermatitis. At the same time, chitosan can also adsorb heavy metals from makeup residues and environmental pollution, reduce skin burden, and lower the risk of allergies. In the realm of hair care and styling, chitosan can neutralize static electricity in shampoos and mitigate friction-induced damage to hair strands. (Suter et al., 2025); Combining with keratin to form a protective film, enhancing hair gloss. In addition, chitosan can also be used in styling products such as hair gel and mousse to provide long-lasting styling effects for hair (Tan et al., 2025). At present, chitosan has been widely used in facial mask, lotion, essence, hair gel, shampoo and other cosmetics. For example, the anti-aging repair essence launched by ChitoCare Beauty, clinical data shows that after 6 weeks of use, the skin's moisturizing degree increases by 16%, and its firmness increases by 22% (Selim et al., 2025).

With the continuous development of the cosmetics industry, the application fields of chitosan are also constantly expanding and deepening. Currently, the cutting-edge exploration of chitosan in the cosmetics field mainly focuses on technological modification, environmental trends, and cross disciplinary collaboration (Zhang et al., 2025g). In terms of technical modification, researchers have improved the water solubility and biological activity of chitosan through chemical modification methods such as carboxymethylation and oligomerization. For example, carboxymethyl chitosan is soluble at neutral pH, which enhances the stability of lotion and can be used as a non-irritating preservative to replace traditional chemical preservatives (Zhou et al., 2025a). Oligochitosan, due to its reduced molecular weight, enhances skin permeability and promotes the absorption of active ingredients. In line with the prevailing emphasis on environmental consciousness in the cosmetics sector, chitosan, as a natural, renewable, and non-toxic green ingredient, is increasingly favored by an expanding number of cosmetic brands. Its biodegradability makes chitosan facial mask substrate and other products completely degradable after use, reducing environmental pollution (Zhao et al., 2025). In addition, the sustainable procurement of chitosan is also in line with the current pursuit of environmental protection and sustainable development in the cosmetics industry. In terms of cross disciplinary collaboration, the application of chitosan in the cosmetics industry is gradually integrating with fields such as medical beauty and health products. For instance, postoperative repair products integrated with chitosan's reparative properties have obtained dermatological clinical certification, thereby expanding their reach within the medical aesthetics market. (Panda et al., 2025). At the same time, the combination of oral chitosan health products and external cosmetics also achieves a dual skincare effect of promoting metabolism through oral administration and strengthening the barrier through external application.

Conclusion and Prospects

Chitosan is a natural polysaccharide polymer material, whose chemical structure is composed of alternating β-(1, 4)-glycosidic bonds between D-glucosamine and N-acetyl-D-glucosamine, forming a linear semi crystalline structure. The free amino groups in D-glucosamine units endow chitosan with cationic properties and pH responsiveness, allowing it to dissolve under acidic conditions and carry a positive charge, thereby interacting with negatively charged substances. The degree of acetylation and molecular weight distribution of chitosan are key parameters that regulate its physical and chemical properties. Chitosan with a high degree of deacetylation exhibits superior solubility and biological activity, attributed to its high density of free amino groups. High molecular weight chitosan, on the other hand, exhibits excellent mechanical properties and film-forming properties due to strong hydrogen bonding between molecular chains. The structural differences of chitosan from different sources are significant. Chitosan from crustaceans is characterized by high molecular weight and low acetylation degree, making it suitable for the preparation of high-strength materials; Fungal derived chitosan exhibits unique advantages in the biomedical field due to its narrow molecular weight distribution and strong controllability of acetylation degree; Insect derived chitosan has become a research hotspot in functional foods due to its antioxidant and anti-inflammatory activities. The functional characteristics of chitosan include biocompatibility, degradability, antibacterial activity, immunomodulatory activity, and antioxidant properties, which make chitosan widely applicable in biomedical, agricultural, food, environmental protection, and cosmetic fields.

In the future, the research and application of chitosan will develop towards precision and multifunctionality. At the technical level, structural modification technology will be the key to breaking through existing performance limitations. Through chemical modifications such as quaternization, carboxymethylation, and nanocomposite technology, chitosan can be endowed with new functions such as targeted antibacterial and intelligent responsive release when combined with graphene oxide and carbon nanotubes. Interdisciplinary integration will provide new driving forces for the innovative application of chitosan. A compelling illustration is that in the biomedical field, the integration of chitosan with 3D printing technology enables the customization of personalized tissue engineering scaffolds. In addition, the potential of chitosan in environmental protection urgently needs to be explored, such as developing biodegradable packaging materials to replace traditional plastics, or constructing efficient heavy metal pollution remediation systems. With the increasing preference of consumers for natural and safe products, as well as the global pursuit of carbon neutrality goals, chitosan, as a biobased material, will play a greater role in replacing traditional petrochemical products and promote the transformation of related industries towards green and efficient directions.

Although chitosan has broad application prospects, its large-scale application still faces multiple challenges. The primary issue is the stability and quality control of raw material supply. Chitosan, derived from crustaceans, relies on fishery waste and is subject to seasonal and regional limitations, resulting in fluctuations in raw material prices and differences in quality. Although fungal fermentation can achieve industrial production, optimizing strain selection and fermentation processes to reduce costs remains a technical bottleneck. In addition, the consistency of chitosan performance needs to be ensured through standardized characterization methods to address the issue of performance differences between different batches of products. The safety evaluation of functional modification constitutes another key challenge, given that chemical modification may introduce potential toxicological risks, and the resultant biological safety must be validated via rigorous toxicological assays. Market promotion and consumer education cannot be ignored. Although the natural properties of chitosan conform to the trend of green consumption, its complex functional mechanism and diverse application scenarios still need to be promoted through science popularization to enhance public awareness. In addition, establishing a cross-industry cooperation platform to promote collaborative innovation among industry, academia, research, and application will be the only way to break through technological barriers and accelerate the industrialization of chitosan.

Acknowledgments

This chapter was supported by Talent Recruitment Program of Shaanxi University Technology [grant numbe: SLGRCQD037, X20250004], and the Shaanxi Provincial Science and Technology Department Project [grant numbe: 2024JC-YBQN-0835].

References

Abdullahi, A., Korumilli, T., & Rao, K. J. (2025). Development and characterization of sodium tripolyphosphate cross-linked chitosan/guar gum composite films reinforced with bentonite clay for food packaging application. *Next Materials, 9*, 100950.

Afrose, N., Chakraborty, R., & Rajendran, K. (2025). Plasma-polymerized chitosan/agarose/chitosan-agarose composite coatings for antibacterial and antibiofilm wound dressings: A review. *International Journal of Biological Macromolecules, 320*, 145873.

Afsharikhah, S., Mavaei, M., Babaei, P., Rashki, S., Ghobadi, S., & Bamneshin, K. (2025). Targeting bacterial biofilm-related genes of Pseudomonas aeruginosa with chitosan-loaded ZnO nanocomposite. *Carbohydrate Polymer Technologies and Applications, 11*, 100899. https://doi.org/https://doi.org/10.1016/j.carpta.2025.100899.

Agustiany, E. A., Nawawi, D. S., Fatriasari, W., Wahit, M. U., Vahabi, H., Kayla, D. S., & Hua, L. S. (2025). Mechanical, morphological, thermal, and fire-retardant properties of sustainable chitosan-lignin based bioplastics. *International Journal of Biological Macromolecules, 306*, 141445. https://doi.org/https://doi.org/10.1016/j.ijbiomac.2025.141445.

Al-Amrah, H., Sharawi, Z. W., Zubair, M. A., Alatawy, M., Mohamed, A. S., Othman, G., El-Magd, M. A. (2025). Impact of propolis and/ or chitosan nanoparticles on H1N1 influenza A virus: In vitro study. *Journal of Radiation Research and Applied Sciences, 18*(2), 101537. https://doi.org/https://doi.org/10.1016/j.jrras.2025.101537.

Alamrani, N. A. (2025). Development of an expanded graphite/locust bean gum/chitosan composite for the adsorption of Lead ions from an aqueous medium. *International Journal of Biological Macromolecules, 318*, 145023. https://doi.org/https://doi.org/10.1016/j.ijbiomac.2025.145023.

Alzahrani, S. O., Alisaac, A., Alsahag, M., Aldosari, F. M., Alshammari, N. M., Alhalafi, M. H., El-Metwaly, N. M. (2025). Green synthesis, characterization, phytochemical analyses, and antibacterial evaluation of Ag/TiO2 and Ag/TiO2-chitosan nanocomposites derived from M. chamomilla. *International Journal of Biological Macromolecules, 312*, 144220. https://doi.org/https://doi.org/10.1016/j.ijbiomac.2025.144220.

Aman, A. F., Ghazy, O. A., Saleh, H., Ali, I. A., Ahmed, N. H., Rizk, S. A., & Ali, Z. I. (2025). Enhanced biomedical performance of magnetic Fe_3O_4@chitosan nanocomposites: A comparative study of synthesis methods for hyperthermia and chemotherapy applications. *JCIS Open, 19*, 100142. https://doi.org/https://doi.org/10.1016/j.jciso.2025.100142.

Apăvăloaiei, I., Nacu, I., Cojocaru, F.-D., Balan, V., Bercea, M., Niţă, L. E., & Vereştiuc, L. (2025). Effect of chitosan on 3D printed scaffolds with gelatin–hyaluronic acid, hydroxyapatite and magnetic nanoparticles for bone tissues defects repair. *Reactive and Functional Polymers, 216*, 106422. https://doi.org/https://doi.org/10.1016/j.reactfunctpolym.2025.106422.

Ashiq, A., Das, S. C., Biswas, S., Akter, R., Masum, M. A., & Bedoura, S. (2025). Accelerating wound healing: Unveiling a synergistic neem-chitosan composite for rapid tissue regeneration. *Next Research, 2*(3), 100541. https://doi.org/https://doi.org/10.1016/j.nexres.2025.100541.

Athanasopoulou, E., Mitsoula, S., Dimitroglou, A., & Tsironi, T. (2025). Chitosan-based biodegradable coatings for securing nutrients in extruded fish feed pellets. *Aquaculture and Fisheries*. https://doi.org/https://doi.org/10.1016/j.aaf.2025.04.001.

Attri, R., Singh, A., Kaur, J., Singh, G., Saxena, A., Khalid, M., & Albakri, G. S. (2025). Sustainable carbon quantum dots–chitosan hydrogel composite for corrosion protection of mild steel in 1 M sulfuric acid. *International Journal of Biological Macromolecules*, 146391. https://doi.org/https://doi.org/10.1016/j.ijbiomac.2025.146391.

Bahamonde-Duarte, J., Briceño, S., Vizuete, K., Debut, A., Borrero-González, L. J., & González, G. (2025). Fluorescent Chitosan-Europium hydroxyapatite beads for drug delivery. *Carbon Trends, 20*, 100513. https://doi.org/https://doi.org/10.1016/j.cartre.2025.100513.

Bahndral, A., Shams, R., Dash, K. K., Chaudhary, P., Mukarram Shaikh, A., & Béla, K. (2025). Microwave assisted extraction of chitosan from Agaricus bisporus: techno-functional and microstructural properties. *Carbohydrate Polymer Technologies and Applications, 9*, 100730. https://doi.org/https://doi.org/10.1016/j.carpta.2025.100730.

Basso, L., Vuillet, C., Rahbé, Y., David, L., Charlot, A., & Sudre, G. (2025). Chitin-binding protein behavior at chitosan interface studied by quartz crystal microbalance with dissipation monitoring (QCM-D): Binding quantification, orientation and affinity constants. *Colloids and Surfaces B: Biointerfaces, 252*, 114650. https://doi.org/https://doi.org/10.1016/j.colsurfb.2025.114650.

Bernabé Vírseda, I., de la Orden, M. U., & Martínez Urreaga, J. (2025). Effects of chitosan thin barrier layers on the oxygen permeability and optical properties of poly(lactic acid) and poly(3-hydroxybutyrate-co-3-hydroxyvalerate) multilayers. *Carbohydrate Polymer Technologies and Applications, 11*, 100918. https://doi.org/https://doi.org/10.1016/j.carpta.2025.100918.

Binder, L., de Sousa Santos, F., & da Conceição, T. F. (2025). Sustainable smart coating of chitosan, halloysite nanotubes and phenolic acids for corrosion protection of Al alloy *Materials Chemistry and Physics, 345*, 131240. https://doi.org/https://doi.org/10.1016/j.matchemphys.2025.131240.

Biswas, U. K., Bose, A., Ghosh, B., & Sharma, S. (2025). An insight into chemically modified chitosan and their biological, pharmaceutical, and medical applications: A review. *International Journal of Biological Macromolecules, 303*, 140612. https://doi.org/https://doi.org/10.1016/j.ijbiomac.2025.140612.

Carbone, C., Navarro-Arrebola, I., Liggieri, L., Ortega, F., Rubio, R. G., & Guzmán, E. (2025). Interfacial dilational rheology of chitosan-silica nanocomposite films at the aqueous dispersion/air interface. *Journal of Molecular Liquids, 425*, 127273. https://doi.org/https://doi.org/10.1016/j.molliq.2025.127273.

Castro, V., Fernández, M. L. G., Negri, R. M., Lizarraga, L., Cánneva, A., Aráoz, B., & Saleh Medina, L. M. (2025). Waste based chitosan membranes for dye removal. *International Journal of Biological Macromolecules, 313*, 143595. https://doi.org/https://doi.org/10.1016/j.ijbiomac.2025.143595.

Çatal, T., Bayrak, A., Çanlıoğlu, A., Çakmur, B. B., Tıngır, M., Kural, A., Duramaz, A. (2025). The role of chitosan in tissue healing after primary repair of quadriceps tendon ruptures: Experimental animal model. *The Knee, 55*, 193-205. https://doi.org/https://doi.org/10.1016/j.knee.2025.04.018.

Charles, A. P. R., Rajasekaran, B., Awasti, N., Choudhary, P., Khanashyam, A. C., Majumder, K., . . . Jin, T. Z. (2025). Emerging chitosan systems incorporated with polyphenols: Their applications in intelligent packaging, active packaging, and nutraceutical systems – A comprehensive review. *International Journal of Biological Macromolecules, 308*, 142714. https://doi.org/https://doi.org/10.1016/j.ijbiomac.2025.142714.

Chelminiak-Dudkiewicz, D., Dlugaszewska, J., Wujak, M., Smolarkiewicz-Wyczachowski, A., Mylkie, K., Machacek, M., & Ziegler-Borowska, M. (2025). Chitosan- and levan-based wound dressings crosslinked with dialdehyde levan: Structural and biological stability under climatic aging. *International Journal of Biological Macromolecules, 321*, 146207. https://doi.org/https://doi.org/10.1016/j.ijbiomac.2025.146207.

Chevrier, A., Wang, D., Ladd, E., Blais, J.-S., & Lavertu, M. (2025). Chitosan powders and Ortho-R, a chitosan-containing finished product, are stable long term when stored in proper conditions. *Carbohydrate Polymer Technologies and Applications, 10*, 100777. https://doi.org/https://doi.org/10.1016/j.carpta.2025.100777.

Choi, H., Yanagita, M., Teramoto, Y., & Takano, T. (2025). Preparation of hemin-bound chitosan flakes and their dye decolorizing ability. *International Journal of Biological Macromolecules, 307*, 141952. https://doi.org/https://doi.org/10.1016/j.ijbiomac.2025.141952.

Chovancová, F., Motiei, M., Šišoláková, I., Urbánek, M., Shepa, J., Fei, H., Oriňaková, R. (2025). Synthesis of nickel dopped chitosan nanoparticles as a novel platform for electrochemical insulin detection. *Biosensors and Bioelectronics: X, 25*, 100624. https://doi.org/https://doi.org/10.1016/j.biosx.2025.100624.

Cibotaru, S., Anisiei, A., Platon, V.-M., Rosca, I., Sandu, I.-A., Coman, C.-G., Marin, L. (2025). Imino-quaternized chitosan/chitosan nanofibers loaded with norfloxacin as potential bandages for wound healing. *International Journal of Biological Macromolecules, 314*, 144304. https://doi.org/https://doi.org/10.1016/j.ijbiomac.2025.144304.

Cignolo, D., Rizzi, V., Fini, P., Petrella, A., Milanese, C., Cosma, P., & Gubitosa, J. (2025). Composite Chitosan-Biochar-based sponges for the removal of Direct Blue-78 from water: Physical and chemical parameters affecting the adsorption process. *Colloids and Surfaces A: Physicochemical and Engineering Aspects, 724*, 137518. https://doi.org/https://doi.org/10.1016/j.colsurfa.2025.137518.

Ćirić, A., Martić, N., Vukmirović, S., Andrejić-Višnjić, B., Janković, T., Rašković, A., & Djekic, L. (2025). Chitosan/xanthan gum polyelectrolyte complex microparticles for enhanced oral delivery of ibuprofen and escin: pharmacokinetic and safety assessment in Wistar rats. *International Journal of Biological Macromolecules, 315*, 144383. https://doi.org/https://doi.org/10.1016/j.ijbiomac.2025.144383.

Coer, O. E., Davidson, B. L., Boardman, B. M., & Peters, G. M. (2025). Modulating Thermal Stability and Flexibility in Chitosan Films with Neutral Polyol-Boric Acid

Complexes. *Biomacromolecules, 26*(7), 4174-4183. https://doi.org/https://doi.org/10.1021/acs.biomac.5c00177.

Crema, N. M., Castro, L. E. N. d., Alves, H. J., & Ribeiro, L. F. (2025). Impact of incorporating chitosan into starch films: analysis of physicochemical properties and degradability in soil. *Food Chemistry Advances, 8*, 101059. https://doi.org/https://doi.org/10.1016/j.focha.2025.101059.

da Silva Bruckmann, F., Gonçalves, J. O., Silva, L. F. O., Oliveira, M. L. S., Dotto, G. L., & Rhoden, C. R. B. (2025). Chitosan-based adsorbents for wastewater treatment: A comprehensive review. *International Journal of Biological Macromolecules, 309*, 143173. https://doi.org/https://doi.org/10.1016/j.ijbiomac.2025.143173.

da Silva Ponciano, C., Gonzaga, F. C., & de Oliveira, C. P. (2025). Smart packaging based on chitosan acting as indicators of changes in food: A technological prospecting review. *International Journal of Biological Macromolecules, 319*, 145722. https://doi.org/https://doi.org/10.1016/j.ijbiomac.2025.145722.

Davydova, V. N., Komleva, D. D., Volod'ko, A. V., Sidorin, E. V., Pimenova, E. A., Kitan, S. A., Yermak, I. M. (2025). Comparative analysis of physicochemical properties and biological activities of chitosans from squid gladius and crab shell. *Carbohydrate Polymers, 365*, 123812. https://doi.org/https://doi.org/10.1016/j.carbpol.2025.123812.

Dejene, B. K., & Abtew, M. A. (2025). Chitosan/zinc oxide (ZnO) nanocomposites: A critical review of emerging multifunctional applications in food preservation and biomedical systems. *International Journal of Biological Macromolecules, 316*, 144773. https://doi.org/https://doi.org/10.1016/j.ijbiomac.2025.144773.

Demircivi, P., Buğdaycı, M., & Kanmaz, N. (2025). Effects of ball-milling time and chitosan/boron carbide ratio on tetracycline adsorption performance. *Journal of the Indian Chemical Society*, 101972. https://doi.org/https://doi.org/10.1016/j.jics.2025.101972.

Deyab, M. A., Mohsen, Q., & El-Shamy, O. A. A. (2025). Toward protection of desalination steel structure from corrosion in harsh marine environment by magnetite-chitosan epoxy nanocomposite. *International Journal of Biological Macromolecules*, 146384. https://doi.org/https://doi.org/10.1016/j.ijbiomac.2025.146384.

Duymaz, D., Kebabci, A. O., & Kizilel, S. (2025). Harnessing the immunomodulatory potential of chitosan and its derivatives for advanced biomedical applications. *International Journal of Biological Macromolecules, 307*, 142055. https://doi.org/https://doi.org/10.1016/j.ijbiomac.2025.142055.

El-Saadony, M. T., Saad, A. M., Alkafaas, S. S., Dladla, M., Ghosh, S., Elkafas, S. S., Hafez, W., Ezzat, S. M., Khedr, S. A., Hussien, A. M., Fahmy, M. A., Elesawi, I. E., Salem, H. M., Mohammed, D. M., Abd El-Mageed, T. A., Ahmed, A. E., Mosa, W. F. A., El-Tarabily, M. K., AbuQamar, S. F., El-Tarabily, K. A., 2025. Chitosan, derivatives, and its nanoparticles: Preparation, physicochemical properties, biological activities, and biomedical applications–A comprehensive review. *International Journal of Biological Macromolecules* 313, 142832.

Edo, G. I., Ndudi, W., Ali, A. B. M., Yousif, E., Zainulabdeen, K., Akpoghelie, P. O., Alamiery, A. A. (2025). Chitosan: An overview of its properties, solubility, functional technologies, food and health applications. *Carbohydrate Research, 550*, 109409. https://doi.org/https://doi.org/10.1016/j.carres.2025.109409.

Esteve-Redondo, P., Heras-Mozos, R., Hernández, R., Gavara, R., & Hernández-Muñoz, P. (2025). Optimisation of stimuli-responsive films based on dynamic α,β-unsaturated imines from chitosan and trans-2-hexenal to enhance the antimicrobial acid-response. *International Journal of Biological Macromolecules, 310*, 143303. https://doi.org/https://doi.org/10.1016/j.ijbiomac.2025.143303.

Fan, L., Peng, R., Liu, X., Li, Z., Wang, Q., & Peng, S. (2025). Superior performance of a novel chitosan-based flocculant in sludge treatment: Synthesis, chelating mechanism, and engineering application by interconnected porous structure. *Journal of Environmental Chemical Engineering, 13*(5), 117841. https://doi.org/https://doi.org/10.1016/j.jece.2025.117841.

Fang, J., Wang, J., Han, Y., Qi, D., & Chen, T. (2025). Bio-based waterproof and oil-repellent coatings for paper packaging: Synergistic effect of oleyl alcohol and chitosan on performance enhancement. *Progress in Organic Coatings, 208*, 109526. https://doi.org/https://doi.org/10.1016/j.porgcoat.2025.109526.

Fusco, A., Guarnieri, A., Scieuzo, C., Triunfo, M., Salvia, R., Donnarumma, G., & Falabella, P. (2025). Hermetia illucens-Derived Chitosan: A Promising Immunomodulatory Agent for Applications in Biomedical Fields. *Biomacromolecules, 26*(5), 3224-3233. https://doi.org/https://doi.org/10.1021/acs.biomac.5c00362.

Gao, C., Xing, Z., Chen, Y., Meng, L., & Tang, X. (2025). Improve the chitosan particle-stabilized oil-water interface by dual reinforcement and its effect on the structure and properties of emulsion. *Food Chemistry, 485*, 144500. https://doi.org/https://doi.org/10.1016/j.foodchem.2025.144500.

Gedik, F., Abdullah, T., Rüzgar, Ş. T., Keteci, F., Bakan, B., Cosut, B., & Dizman, C. (2025). Chitosan-based enzyme-immobilized nanocomposite cryogels with controlled biodegradation profile. *International Journal of Biological Macromolecules, 318*, 145275. https://doi.org/https://doi.org/10.1016/j.ijbiomac.2025.145275.

Ghattas, M., Chevrier, A., Wang, D., Alameh, M.-G., & Lavertu, M. (2025). Reducing endotoxin contamination in chitosan: An optimized purification method for biomedical applications. *Carbohydrate Polymer Technologies and Applications, 10*, 100797. https://doi.org/https://doi.org/10.1016/j.carpta.2025.100797.

Ghorai, P., Jana, B., Ghorai, S., & Ganguly, J. (2025). Synthetic strategy for multimodal –NH2 functionalized chitosan-based materials towards sustainable evolution. *Carbohydrate Research, 552*, 109482. https://doi.org/https://doi.org/10.1016/j.carres.2025.109482.

Ghosh, M., Viswaroopan, N., Kshirsagar, S. M., Khan, J., Mohiuddin, S., Srivastava, R. K., Banga, A. K. (2025). Sustained delivery of 4-phenylbutyric acid via chitosan nanoparticles in foam for decontamination and treatment of lewisite-mediated skin injury. *International Journal of Pharmaceutics, 682*, 125928. https://doi.org/https://doi.org/10.1016/j.ijpharm.2025.125928.

Ghysels, S., Van Poucke, C., Ronsse, F., Mangelinckx, S., & Stevens, C. V. (2025). Uncovering the blind spot in the analysis of N-alkylidene and N-alkyl chitosan derivatives through Py-GC-MS/FID. *Carbohydrate Research, 553*, 109503. https://doi.org/https://doi.org/10.1016/j.carres.2025.109503.

Gibeaux, C., Delafosse, D., Reinert, L., Fontvieille, D., & Duclaux, L. (2025). Chitosan hydrogels as passive samplers for the analysis of norovirus in various types of water. *Journal of Water Process Engineering, 76*, 108032. https://doi.org/https://doi.org/10.1016/j.jwpe.2025.108032.

Gopalakrishnan, G., Deepa, P. M., Rajasekhar, R., Vergis, J., Bipin, K. C., Rathish, R. L., & Janus, A. (2025). In vitro antiviral activity of chitosan and curcumin-loaded chitosan nanoparticles against virulent Newcastle disease virus isolates. *Microbial Pathogenesis, 207*, 107870. https://doi.org/https://doi.org/10.1016/j.micpath.2025.107870.

Gunasekaran, H. A., Pitchai, A., & Ramasamy, P. (2025). Synthesis of sulfated chitosan from cuttlebone of Sepia pharaonis (Ehrenberg, 1831) and its capacity to inhibit blood clotting. *Bioactive Carbohydrates and Dietary Fibre, 33*, 100470. https://doi.org/https://doi.org/10.1016/j.bcdf.2025.100470.

He, X., Zhang, Z., Xuan, X., Tan, T., Sun, J., Wang, B., Chen, H. (2025). Structure and properties of chitosan plasticized with hydrophobic short-chain fatty acid cosolvent. *International Journal of Biological Macromolecules, 300*, 140250. https://doi.org/https://doi.org/10.1016/j.ijbiomac.2025.140250.

He, Y., & Wu, S. (2025). Chitosan–garlic extract coating improves germination, antioxidant defense, and antifungal protection of soybean seeds during summer storage. *Journal of Stored Products Research, 114*, 102724. https://doi.org/https://doi.org/10.1016/j.jspr.2025.102724.

Heidari, F., Dianati Tilaki, G. A., Kooch, Y., & Abdollahi, M. (2025). Improving soil function properties in semi-arid regions using modified-chitosan and biochar. *Journal of Environmental Management, 390*, 126334. https://doi.org/https://doi.org/10.1016/j.jenvman.2025.126334.

Hevira, L., Ighalo, J. O., Sondari, D., & Hanifah, A. S. (2025). Evaluating the efficiency and scalability of chitosan from black soldier fly in removing Congo red from wastewater. *Bioresource Technology Reports, 30*, 102132. https://doi.org/https://doi.org/10.1016/j.biteb.2025.102132.

Hsu, C.-Y., Ahmed, A. Y., Ullah, M. I., Ballal, S., Singh, A., Kavitha, V., Hashemzadeh, A. (2025). MUC1 aptamer-functionalized chitosan nanoparticles: Targeted delivery platforms for synergistic cancer therapy and multidrug resistance overcoming. *International Journal of Biological Macromolecules, 319*, 145514. https://doi.org/https://doi.org/10.1016/j.ijbiomac.2025.145514.

Hussain, M. S., Alfawaz Altamimi, A. S., Afzal, M., Babu, M. A., Goyal, K., R, R., Balaraman, A. K. (2025). Chitosan-based hydrogels: Revolutionizing corneal wound healing with antibacterial and regenerative properties. *Experimental Eye Research, 258*, 110503. https://doi.org/https://doi.org/10.1016/j.exer.2025.110503.

Ibrahim, A. G., Elgammal, W. E., El-Sayed Ebead, E., Ismail, M. F., Sharaf, M. H., & Aboelnaga, A. (2025). Functionalized chitosan with new acrylohydrazide derivative: Synthesis, characterization, DFT calculations, antimicrobial and antiproliferative activities. *Journal of Molecular Structure, 1341*, 142607. https://doi.org/https://doi.org/10.1016/j.molstruc.2025.142607.

Iftime, M. M., Ailiesei, G. L., Morariu, S., Sandu, A.-I., Rambu, C. M., & Marin, L. (2025). Functional biocompatible chitosan hydrogels crosslinked with a vanillin isomer: Controlled antioxidant, antimicrobial, and self-healing properties. *International Journal of Biological Macromolecules, 319*, 145411. https://doi.org/https://doi.org/10.1016/j.ijbiomac.2025.145411.

Im, H. Y., Kim, J. S., & Lim, J. K. (2025). Preparation of gas-filled self-floating chitosan capsules and their application in the removal of methyl orange dye. *Reactive and Functional Polymers, 213*, 106269. https://doi.org/https://doi.org/10.1016/j.reactfunctpolym.2025.106269.

Iqbal, F., Upadhyay, A., Tehrani, R., & Rowles, L. S. (2025). Evaluating environmental and economic sustainability of engineered chitosan particles for water treatment. *Cleaner Materials, 16*, 100319. https://doi.org/https://doi.org/10.1016/j.clema.2025.100319.

Jin, J., Ren, Y., Zhou, H., Hou, Y., Jin, P., Zheng, Y., & Wu, Z. (2025). Multi-factor protective chitosan/polyvinyl alcohol aerogel antibacterial buffer packaging. *Chemical Engineering Journal, 516*, 164097. https://doi.org/https://doi.org/10.1016/j.cej.2025.164097.

Jogaiah, S., Mujtaba, A. G., Mujtaba, M., Archana, De Britto, S., Geetha, N., Shetty, H. S. (2025). Chitosan-metal and metal oxide nanocomposites for active and intelligent food packaging; a comprehensive review of emerging trends and associated challenges. *Carbohydrate Polymers, 357*, 123459. https://doi.org/https://doi.org/10.1016/j.carbpol.2025.123459.

Kedves, A., & Kónya, Z. (2025). Enhancing granule formation: The role of chitosan and chitosan nanoparticles in microalgal-bacterial granular sludge development. *Environmental Technology & Innovation, 39*, 104240. https://doi.org/https://doi.org/10.1016/j.eti.2025.104240.

Khainskaya, K., Hileuskaya, K., Nikalaichuk, V., Ladutska, A., Akhmedov, O., Abrekova, N., Odonchimeg, M. (2025). Chitosan-gallic acid conjugate with enhanced functional properties and synergistic wound healing effect. *Carbohydrate Research, 553*, 109496. https://doi.org/https://doi.org/10.1016/j.carres.2025.109496.

Khani, F. M. R., Naseri, M., Babaei, S., Hosseini, S. M. H., & Valizadeh, S. (2025). Cumin essential oil-infused chitosan-PVA composite coatings: An innovative approach for developing antimicrobial, antioxidant and functionalized polyethylene films. *International Journal of Biological Macromolecules*, 145270. https://doi.org/https://doi.org/10.1016/j.ijbiomac.2025.145270.

Khodavandegar, S., Zare, R., & Fatehi, P. (2025). Carboxymethylated lignin incorporated chitosan aerogel as thermal insulator. *Industrial Crops and Products, 234*, 121540. https://doi.org/https://doi.org/10.1016/j.indcrop.2025.121540.

Kiaee Nasab Abkenar, S. M., & Dehnavi, S. M. (2025). Laccase immobilization onto MIL–100 (Fe) and chitosan nanofibers: Fabrication, characterization, and investigation of the potential for bisphenol A removal. *Journal of Water Process Engineering, 74*, 107827. https://doi.org/https://doi.org/10.1016/j.jwpe.2025.107827.

Kobisy, A. S., Guan, F., Duan, J., Tawfik, S. M., Guo, D., Wang, Y., & Salgar-Chaparro, S. J. (2025). Mitigation of marine microbial corrosion using natural products based on chitosan polymer. *International Journal of Biological Macromolecules*, 145429. https://doi.org/https://doi.org/10.1016/j.ijbiomac.2025.145429.

Kumar, D., Khullar, P., Jain, V., Prakash, R., & Rai, B. (2025). Investigating the corrosion inhibition mechanism of chitosan for steels in acidic environments: Experimental and computational studies. *Journal of Molecular Structure, 1337*, 142057. https://doi.org/https://doi.org/10.1016/j.molstruc.2025.142057.

Li, J., Fang, Y., Li, Q., Zeng, C., Jiang, Y., Kong, W., & Zhu, M. (2025). Dual-mode regulatory cross-linked chitosan nanocapsules for pH-responsive and controllable sustained release in textiles. *International Journal of Biological Macromolecules, 314*, 144109. https://doi.org/https://doi.org/10.1016/j.ijbiomac.2025.144109.

Li, R., He, X., Yang, L., Xiang, X., Jiang, S., Lin, L., & Lu, J. (2025). From Chinemys reevesii shell to trauma care: Preparation and characterization of chitosan-tortoiseshell collagen composite sponge with hemostasis property. *International Journal of Biological Macromolecules, 321*, 146188. https://doi.org/https://doi.org/10.1016/j.ijbiomac.2025.146188.

Li, Y., Li, X., Zhu, L., Liu, T., & Huang, L. (2025). Chitosan-based biomaterials for bone tissue engineering. *International Journal of Biological Macromolecules, 304*, 140923. https://doi.org/https://doi.org/10.1016/j.ijbiomac.2025.140923.

Lin, Y., Lun, W., Bai, Y., Che, Q., Cao, H., Guo, J., & Su, Z. (2025). Chitosan lactate mitigates metabolic-associated fatty liver disease via TGR5-mediated anti-inflammatory mechanisms and the gut-hepatic-bile acid axis. *Journal of Functional Foods, 131*, 106946. https://doi.org/https://doi.org/10.1016/j.jff.2025.106946.

Liu, C., Crini, G., Lichtfouse, E., Wilson, L. D., Picos-Corrales, L. A., Balasubramanian, P., & Li, F. (2025). Chitosan-based materials for emerging contaminants removal: Bibliometric analysis, research progress, and directions. *Journal of Water Process Engineering, 71*, 107327. https://doi.org/https://doi.org/10.1016/j.jwpe.2025.107327.

Liu, Y., Qu, M., Cao, K., & Wang, D. (2025). Preservation properties of Toona sinensis essential oil-chitosan composite films: A comprehensive study. *Applied Food Research, 5*(2), 101194. https://doi.org/https://doi.org/10.1016/j.afres.2025.101194.

Ma, X., Lv, J., Fu, J., Hao, H., Bi, J., Hou, H., & Zhang, G. (2025). Characterization of phenethyl isothiocyanate nanoemulsion loaded pectin/chitosan composite film and application in beef preservation. *Food Chemistry, 490*, 145030. https://doi.org/https://doi.org/10.1016/j.foodchem.2025.145030.

Ma, Y., Tian, R., Song, Y., Liu, M., Li, J., Bi, Z., Li, Y. (2025). Schiff base crosslinking based dialdehyde-chitosan/gelatin/tea polyphenol electrospun nanofibrous membranes with excellent antibacterial and antioxidant activity for food packaging. *Food Packaging and Shelf Life, 50*, 101568. https://doi.org/https://doi.org/10.1016/j.fpsl.2025.101568.

Mahaninia, M. H., & Yan, N. (2025). Chitosan-based partial Vitrimer for making eco-friendly multifunctional ramie fabric reinforced composite. *International Journal of Biological Macromolecules, 309*, 142979. https://doi.org/https://doi.org/10.1016/j.ijbiomac.2025.142979.

Mandal, S., Hazra, C., Joshi, N., Kumar, P., Sen, R., Das, S., & Das, K. (2025). Development and performance evaluation of chitosan hydrogel based ZnO/CuO antibacterial and antialgal nanocoating for the control of biofouling in cultural heritage stones. *Construction and Building Materials, 489*, 142236. https://doi.org/https://doi.org/10.1016/j.conbuildmat.2025.142236.

Manu, G. P., Narh, D., Asimeng, B., Kwakye, N., Abusuapa, B. A., Appuing, D., Efavi, J. K. (2025). Characterization of Chitosan/ Zeolite drug delivery composite and curcumin release kinetics in a simulated pH environment. *Scientific African, 28*, e02668. https://doi.org/https://doi.org/10.1016/j.sciaf.2025.e02668.

Martingo, M., Baptista-Silva, S., Mesquita, R., Ferreira, J. P., Borges, S., & Pintado, M. (2025). Exploring the potential of mealworm chitosan for hemodialysis applications. *Sustainable Chemistry and Pharmacy, 45*, 102013. https://doi.org/https://doi.org/10.1016/j.scp.2025.102013.

Martingo, M., Baptista-Silva, S., Pintado, M., & Borges, S. (2025). A new vision upon hemodialysis: A shift from synthetic to sustainable chitosan membranes. *Biomedical Engineering Advances, 9*, 100175. https://doi.org/https://doi.org/10.1016/j.bea.2025.100175.

Maury, R., Gardrat, C., Dumortier, C., & Coma, V. (2025). Hydrophobic kaolin-chitosan based coating for Spring frost protection. *Progress in Organic Coatings, 203*, 109169. https://doi.org/https://doi.org/10.1016/j.porgcoat.2025.109169.

Mawale, K. S., Giridhar, P., & Johnson, T. S. (2025). Chitosan: A versatile polymer for enhancing plant bioactive accumulation, managing plant diseases, and advancing food preservation technologies. *International Journal of Biological Macromolecules, 308*, 142081. https://doi.org/https://doi.org/10.1016/j.ijbiomac.2025.142081.

Mehrabi-Khozani, Z., Sarabandi, K., & Rezaei, A. (2025). Co-delivery of hydrophobic curcumin and hydrophilic vitamin C by nanoliposomes decorated by chitosan and gelatin. *Carbohydrate Polymer Technologies and Applications, 11*, 100937. https://doi.org/https://doi.org/10.1016/j.carpta.2025.100937.

Milbreta, U., Andze, L., Zoldners, J., Irbe, I., Skute, M., & Filipova, I. (2025). Exploring the Potential of Locally Sourced Fungal Chitosan for Paper Mechanical Property Enhancement. *Journal of Renewable Materials, 13*(3), 583-597. https://doi.org/https://doi.org/10.32604/jrm.2024.057663.

Mohamed, A. S. A., Abdelaziz, A. A., Abo-Kamar, A. M., Al-Madboly, L. A., El-Maradny, Y. A., & El-Fakharany, E. M. (2025). Multifunctional efficacy of the fabricated chitosan-coated silver nanoparticles as an antiviral agent against SARS-CoV-2: Potent and safe mechanistic insights. *International Journal of Biological Macromolecules, 320*, 145623. https://doi.org/https://doi.org/10.1016/j.ijbiomac.2025.145623.

Mohammadalipour, M., Alihosseini, F., Toloue, E. B., Karbasi, S., & Mohammadalipour, Z. (2025). Evaluation of PHB-chitosan/CNC scaffolds' applicability for bone tissue engineering via MG-63 osteoblastic cell cultivation and osteogenic markers gene expression. *International Journal of Biological Macromolecules, 320*, 146041. https://doi.org/https://doi.org/10.1016/j.ijbiomac.2025.146041.

Moreira, B., Calvet, S., Marques, A., & Martins, G. V. (2026). Chitosan-based molecularly imprinted polymers for point-of-care C-reactive protein detection. *Talanta, 296*, 128465. https://doi.org/https://doi.org/10.1016/j.talanta.2025.128465.

Mu, D., Chen, Q., Gao, B., Xu, H., Chen, W., Wang, K., Ke, S. (2025). Oxidized chitosan: Combining an "Adhesion-and-Kill" antibacterial strategy with immunoregulation and angiogenesis to enhance methicillin-resistant Staphylococcus Aureus-infected wound healing. *International Journal of Biological Macromolecules, 319*, 145664. https://doi.org/https://doi.org/10.1016/j.ijbiomac.2025.145664.

Nabil, Y., Atta, A. H., El-Fattah, W. A., Hafez, H. S., & Elshaarawy, R. F. M. (2025). Progress in chitosan/essential oil/ZnO nanobiocomposites fabrication for wound healing applications: A review. *International Journal of Biological Macromolecules, 318*, 145123. https://doi.org/https://doi.org/10.1016/j.ijbiomac.2025.145123.

Nandhakumar, Ramachandran, I., & Elumalai, P. (2025). Mucoadhesive chitosan-based nano vaccine as promising immersion vaccine against Edwardsiella tarda challenge in Nile tilapia (Oreochromis niloticus). *Veterinary Immunology and Immunopathology, 286*, 110976. https://doi.org/https://doi.org/10.1016/j.vetimm.2025.110976.

Nath, J., Dewan, M., Ghosh, A., Ray, S. S., Orasugh, J. T., Lahiri, B., Adhikari, A. (2025). Chitosan-based adsorbents for remediation of toxic dyes from wastewater: A review on adsorption mechanism, reusability, machine learning based modeling and future perspectives. *International Journal of Biological Macromolecules, 311*, 143388. https://doi.org/https://doi.org/10.1016/j.ijbiomac.2025.143388.

Naveedunissa, S., Meenalotchani, R., Manisha, M., Ankul Singh, S., Nirenjen, S., Anitha, K., Prajapati, B. G. (2025). Advances in chitosan based nanocarriers for targetted wound healing therapies: a review. *Carbohydrate Polymer Technologies and Applications, 11*, 100891. https://doi.org/https://doi.org/10.1016/j.carpta.2025. 100891.

Nguyen, L. H., Nam, N. H., Bach, T. N., Ngoc Nha, T. T., Minh Hong, N. T., Van, T. D., Nam, P. H. (2025). Comprehensive studying the role of coating on structure, magnetic properties, and heating efficiency for chitosan-coated Co0.2Zn0.8Fe2O4 nanoparticles. *Journal of Magnetism and Magnetic Materials, 628*, 173200. https://doi.org/https://doi.org/10.1016/j.jmmm.2025.173200.

Olunusi, S. O., Ramli, N. H., Adam, F., Hossain, E., Kim, T. H., 2025. Chitosan as a sustainable alternative for fresh food packaging: Structural insights, modification strategies, and innovations for commercial viability. *International Journal of Biological Macromolecules* 318, 145065.

Oliveira Júnior, F. D., Pinheiro, A. C., Vicente, A. A., & Cunha, R. L. (2025). Development of chitosan/navy bean extract-based nanocarriers for oral delivery of insulin. *Food Research International, 218*, 116876. https://doi.org/https://doi.org/10.1016/j. foodres.2025.116876.

Osman, M., Sharmin, Z., Suchy, S., Gao, F., Kaminski, A., Mitchell, J. C., Carrilho, M. R. (2025). Bioinspired smart dentin ECM-chitosan hydrogels for dentin-pulp complex regeneration. *Journal of Dentistry, 159*, 105811. https://doi.org/https://doi.org/10. 1016/j.jdent.2025.105811.

Panahimehr, M., Asghari, S., Hosseini, M., & Mojaddami, A. (2025). Pd/$ZnMn_2O_4$/chitosan nanobiocatalyst: A sustainable solution for sunlight-enhanced photocatalytic degradation of Congo red dye. *Journal of Molecular Structure, 1347*, 143339. https://doi.org/https://doi.org/10.1016/j.molstruc.2025.143339.

Panda, T. K., Bollepalli, D., Beji, D. S., Varanasi, K., Ananth, M., Rajarajeswaran, J., & Eswar Neerugatti, K. R. (2025). Beyond Slurry Systems: Chitosan Thin Film Fabrication, Mechanism, and Reusability for Sustainable Textile Effluent

Remediation. *Separation and Purification Technology, 368*, 132936. https://doi.org/https://doi.org/10.1016/j.seppur.2025.132936.

Patel, P., Shoyele, N. O., & Shoyele, S. A. (2025). Preparation and evaluation of meclinertant-loaded PLGA/chitosan nanoparticles as a potential treatment for non-small cell lung cancer using A549 cells. *Journal of Drug Delivery Science and Technology, 107*, 106831. https://doi.org/https://doi.org/10.1016/j.jddst.2025.106831.

Peng, Y., Yu, Y., Su, Z., Zeng, J., Zhong, Y., Chen, Y., Li, M. (2025). Design and synthesis of novel zwitterionic-quaternary ammonium chitosan derivatives of distinct categories: Strategies for enhancing antibacterial, antioxidant, and biocompatible properties. *International Journal of Biological Macromolecules, 309*, 142665. https://doi.org/https://doi.org/10.1016/j.ijbiomac.2025.142665.

Petchsoongsakul, T., Thungphotrakul, N., Jarastrakull, P., Loykulnant, S., Kongkaew, C., Rattanaporn, K., & Prapainainar, P. (2025). Antimicrobial activity enhancement of chitosan nanoparticles/natural rubber composite foam. *International Journal of Biological Macromolecules, 318*, 145337. https://doi.org/https://doi.org/10.1016/j.ijbiomac.2025.145337.

Piccioni, M., Ghignone, S., Peila, R., Vineis, C., Lumini, E., & Tummino, M. L. (2025). Biodegradation pathways in compost-enriched soil of cotton fabrics treated with chitosan and a natural dye: Chemical and biological evaluation. *International Journal of Biological Macromolecules, 313*, 144327. https://doi.org/https://doi.org/10.1016/j.ijbiomac.2025.144327.

Prabhakar, O., & Pavani, N. (2025). Bioengineering of chitosan/citric acid/curcumin nanofiber scaffolds for treatment of spinal cord injury. *International Journal of Biological Macromolecules, 321*, 146143. https://doi.org/https://doi.org/10.1016/j.ijbiomac.2025.146143.

Putra, P. S. U., Adhika, D. R., Genecya, G., Al Madanie, M. S., & Asri, L. A. T. W. (2025). Evaluation of Chitosan-Encapsulated Lemongrass (Cymbopogon citratus) Essential Oil Nanoemulsion for Fruit Edible Coating. *OpenNano, 24*, 100246. https://doi.org/https://doi.org/10.1016/j.onano.2025.100246.

Putro, J. N., Bintang Ilhami, F., Cheng, C.-C., Imanuela, K. J., Mangindaan, D., Irawaty, W., Ismadji, S. (2025). Polyelectrolyte nanoparticles derived from durian seed gum modified with chitosan for selective targeting in cancer cells. *International Journal of Biological Macromolecules, 315*, 144625. https://doi.org/https://doi.org/10.1016/j.ijbiomac.2025.144625.

Rajeswari, M. V., Suresh, A., Thirumalaikumar, E., Subramanian, K., Kaliaperumal, K., Sampath, S., AlSalhi, M. S. (2025). Efficacy of chitosan conjugated bivalent vaccine on immune response and disease resistance against Aeromonas hydrophila, and Aeromonas caviae in goldfish Carassius auratus. *Journal of Drug Delivery Science and Technology, 112*, 107190. https://doi.org/https://doi.org/10.1016/j.jddst.2025.107190.

Remesh, S. D., Sandrasaigaran, P., Remesh, S., Perumal, V., Lip Vun, J. Y., Gandhi, S., & Hasan, H. (2025). Antimicrobial and anti-biofilm activity of crustacean-derived chitosan against Salmonella typhimurium, Staphylococcus aureus and Listeria monocytogenes. *Food Bioscience, 68*, 106697. https://doi.org/https://doi.org/10.1016/j.fbio.2025.106697.

Rezaie, H., Zareie, F., Razzaghi, D., Rezaei, M., & Babaie, A. (2025). Surface functionalization of chitosan nanoparticles with PCL Chains: Preparation of hybrid polyurethane electrospun nanofibers for shape memory-assisted biomedical application. *Surfaces and Interfaces, 71*, 106897. https://doi.org/https://doi.org/10.1016/j.surfin.2025.106897.

Rusho, M. A., Ahmed, T. A., Saleh, L. H., Ghori, S. W., Muniyandy, E., Usanov, S., Akbar, M. (2025). Design and synthesis of decorated palladium nanoparticles on chitosan-tannic acid modified magnetic nanoparticles and evaluation of its catalytic application in the Heck coupling reactions. *Journal of Organometallic Chemistry, 1039*, 123773. https://doi.org/https://doi.org/10.1016/j.jorganchem.2025.123773.

Saaid, A. A., & Ibrahim, B. M. (2025). Modification of chitosan with Kryptofix222 as a novel adsorbent for removing lead and cadmium ions. *International Journal of Biological Macromolecules, 320*, 146124. https://doi.org/https://doi.org/10.1016/j.ijbiomac.2025.146124.

Sayyad, A. H. A. L., Salman, H. I., & Balakit, A. A. (2025). Design and synthesis of a new chitosan derivative for the removal of congo red from aqueous solutions via non-electrostatic interactions. *Desalination and Water Treatment*, 101357. https://doi.org/https://doi.org/10.1016/j.dwt.2025.101357.

Selim, H. M. R. M., Saleh, A., Soliman, N. R., Abdelkhalig, S. M., Mattar, E., Mahmoud, M. M., ... Hamed, A. A. (2025). Myco-fabricated CuO NPs and chitosan-conjugated CuO nanoparticles using endophytic Aspergillus sp. JAWF3: Targeting E. coli outer membrane for enhanced antibacterial and biofilm inhibition. *Inorganic Chemistry Communications, 178*, 114361. https://doi.org/https://doi.org/10.1016/j.inoche.2025.114361.

Serra, J. P., Andonegi, M., Salazar, H., Correia, D. M., de la Caba, K., Guerrero, P., Costa, C. M. (2025). Sustainable soft actuators based on chitosan and choline containing ionic liquids with high bending response. *International Journal of Biological Macromolecules, 320*, 145972. https://doi.org/https://doi.org/10.1016/j.ijbiomac.2025.145972.

Sharma, K., Sonu, S., Singh, P., Ahamad, T., Kaya, S., Katin, K. P., Raizada, P. (2025) Synergy between dual Z-scheme heterostructured Vo-Bi2WO6/Bi2S3/Ag2S@ Chitosan for the effective degradation of fuchsin basic dye. *Journal of the Taiwan Institute of Chemical Engineers, 175*, 106265. https://doi.org/https://doi.org/10.1016/j.jtice.2025.106265.

Sheikhlou, H. G., Sheikhlou, M. G., Saberi, M., & Nojavan, S. (2025). Enhanced electromembrane extraction of acidic drugs using a chitosan-agarose hybrid gel membrane. *Advances in Sample Preparation, 15*, 100196. https://doi.org/https://doi.org/10.1016/j.sampre.2025.100196.

Silva, F. M., de Camargo, R. V., Guedes, D. F. C., de Assis, H. C., Lopes-Olhê, F. C., & da Cruz Filho, A. M. (2025). Enhancing bond strength and sealer penetration of resin-based Sealers: The impact of chitosan/genipin solution in root canal final irrigation. *International Journal of Adhesion and Adhesives, 142*, 104073. https://doi.org/https://doi.org/10.1016/j.ijadhadh.2025.104073.

Smułka, A., Olejnik, A., Zarzeczańska, D., Wcisło, A., Ryl, J., Zieliński, A., & Ossowski, T. (2025). pH-mediated charge tuning of chitosan membrane: a smart approach to control electrochemical selectivity. *Electrochimica Acta, 533*, 146541. https://doi.org/https://doi.org/10.1016/j.electacta.2025.146541.

Soni, M., Deachen, S., Das, S., Yadav, A., Dubey, N. K., & Dwivedy, A. K. (2025). Chitosan-based Illicium verum Hook. f. essential oil nanoemulsion as an edible coating for shelf-life enhancement of custard apple. *International Journal of Biological Macromolecules, 320*, 145564. https://doi.org/https://doi.org/10.1016/j.ijbiomac.2025.145564.

Soussani, F.-E., Hamdan, Y. A., Ait rahou, Y., Rhazi, M., Douira, A., Benkirane, R., & Meddich, A. (2025). Integrated use of chitosan and arbuscular mycorrhizal fungi mitigates Verticillium dahliae infection in Tomato: Comparison with conventional fertilization. *Physiological and Molecular Plant Pathology, 139*, 102787. https://doi.org/https://doi.org/10.1016/j.pmpp.2025.102787.

Srinivasan, P., Pitchai, A., & Ramasamy, P. (2025). Phosphorylated chitosan from the internal bone of Sepia kobiensis (Hoyle, 1885) and their inhibition against oral pathogens. *Carbohydrate Research, 550*, 109413. https://doi.org/https://doi.org/10.1016/j.carres.2025.109413.

Sundari, M. M., Jegatheesan, A., Mohan, R., Arul Raj, M. S., Devi, N. A., Ravikumar, P., Ravichandran, K. (2025). Exploring the bioactive potential of chitosan extracted from Portunus trituberculatus crab shell for multifaceted applications: Antioxidant, antidiabetic, anti-inflammatory and cytotoxic activities. *Materials Letters, 393*, 138581. https://doi.org/https://doi.org/10.1016/j.matlet.2025.138581.

Suter, E., Rutto, H. L., & Mkhize, I. G. (2025). Biodegradable waste-derived cellulose/nylon-6-coated iron-oxide nanocomposite encapsulated with chitosan for enhanced wastewater treatment. *Cleaner Chemical Engineering, 11*, 100190. https://doi.org/https://doi.org/10.1016/j.clce.2025.100190.

Tan, X., Wu, X., Wang, X., Tao, Y., Li, Y., Tang, S., Sheng, Y. (2025). Chitosan-assisted dilute acid pretreatment with enhanced prehydrolysate fermentability and enzyme digestibility of pretreated substrates. *International Journal of Biological Macromolecules, 310*, 143231. https://doi.org/https://doi.org/10.1016/j.ijbiomac.2025.143231.

Thathsarani, N., Khiadani, M., Azhar, M. R., & Zargar, M. (2025). Enhanced removal of microplastic fibres using aluminium and chitosan-based coagulants assisted with microbubble technology. *Journal of Environmental Chemical Engineering, 13*(3), 116780. https://doi.org/https://doi.org/10.1016/j.jece.2025.116780.

Usta, S. N., Avcı, E., Oktay, A. N., & Keskin, C. (2025). Combination of Chitosan Nanoparticles, EDTA, and Irrigation Activation Enhances TGF-β1 Release from Dentin: A Laboratory Study. *Journal of Endodontics*. https://doi.org/https://doi.org/10.1016/j.joen.2025.04.011.

Van Overschelde, P., Vansintjan, P., Portelange, N., Chausson, M., & Weyenberg, W. (2025). Twelve-month efficacy of carboxymethyl-chitosan in refractory knee osteoarthritis: A randomized controlled trial (PIONEER). *Osteoarthritis and Cartilage Open, 7*(2), 100605. https://doi.org/https://doi.org/10.1016/j.ocarto.2025.100605.

Vieira, R. M., Gauthier, E., Widanagamage, G., McKenzie, N., Dunn, K., Moghaddam, L., Brienzo, M. (2025). Uv-blocking and hydrophobicity improvement of chitosan-based film with lignin addition from a pilot scale-up Organosolv process of Banana Pseudostem. *International Journal of Biological Macromolecules, 313*, 144254. https://doi.org/https://doi.org/10.1016/j.ijbiomac.2025.144254.

Vosoughi, P., Naghib, S. M., Jafari, T., & Kangarshahi, B. M. (2025). Chitosan-encapsulated lipid-based nanovesicles for therapeutic applications and tissue engineering: A review. *Carbohydrate Polymer Technologies and Applications, 10*, 100805. https://doi.org/https://doi.org/10.1016/j.carpta.2025.100805.

Wan, J., Yin, M., Wang, F., Huang, G., Pra, I. D., Liang, Y., & Wu, J. (2025). Natural gelatin / chitosan oligosaccharides / emodin eutectogels with multiple antibacterial effects for accelerated infected wound healing. *International Journal of Biological Macromolecules, 320*, 145771. https://doi.org/https://doi.org/10.1016/j.ijbiomac.2025.145771.

Wang, J., Liu, Y., Li, X., & Wu, S. (2025). Peanut seed molds affected by chitosan loaded with different antimicrobial agents. *Journal of Stored Products Research, 112*, 102637. https://doi.org/https://doi.org/10.1016/j.jspr.2025.102637.

Wang, K., Song, Z., Xu, Z., Xi, Y., Cui, Y., & Zhou, H. (2025). Synthesis of quaternized magnetic chitosan and adsorption performance for methyl orange from aqueous solution††Electronic supplementary information (ESI) available. See DOI: https://doi.org/10.1039/d5ra02862k. *RSC Advances, 15*(26), 21121-21132. https://doi.org/https://doi.org/10.1039/d5ra02862k.

Wujcicki, Ł., Mańdok, T., Paprota, D., Sapia, D., Dudek, G., Gołombek, K., & Kluczka, J. (2025). A novel chitosan sorbent enriched with nickel ions: An effective approach for eliminating excess phosphate(V) anions in water. *Water Resources and Industry, 34*, 100301. https://doi.org/https://doi.org/10.1016/j.wri.2025.100301.

Xenou, E. A., Malandrakis, A. A., Katzourakis, V. E., & Chrysikopoulos, C. V. (2025). Interaction of λ-cyhalothrin with chitosan and quartz sand: Attachment, transport, and cotransport in porous media. *Advances in Water Resources, 200*, 104965. https://doi.org/https://doi.org/10.1016/j.advwatres.2025.104965.

Xing, Y., Luoguo, W., Luo, W., Wang, X., Dan, L., Xu, Q., Lv, X. (2025). Effects of chitosan and acetic acid composite coating on the quality and metabolomic characteristics of postharvest pineapple. *Food Chemistry, 484*, 144488. https://doi.org/https://doi.org/10.1016/j.foodchem.2025.144488.

Xu, X., Gao, C., Feng, X., Meng, L., Wang, Z., Zhang, Y., & Tang, X. (2025). Effects of keto acid crosslinking on the structure and properties of chitosan based casted and hot-pressed films. *International Journal of Biological Macromolecules, 308*, 142751. https://doi.org/https://doi.org/10.1016/j.ijbiomac.2025.142751.

Yang, K., Zhu, R., Bao, H., Xu, S., Gao, Y., Xue, Y., Zhao, K. (2025). Chitosan-based nanomaterials: Pioneering a review in veterinary medicine applications. *International Journal of Biological Macromolecules, 320*, 146011. https://doi.org/https://doi.org/10.1016/j.ijbiomac.2025.146011.

Yang, T., Zhang, X., Wang, F., Zhang, L., Yu, D., Zhang, X., Wang, J. (2026). Development of biodegradable multifunctional zein-curcumin-quercetin-chondroitin sulfate nanoparticles with cold plasma-modified chitosan-based active film for tilapia preservation. *Food Hydrocolloids, 170*, 111733. https://doi.org/https://doi.org/10.1016/j.foodhyd.2025.111733.

Yang, X., Chen, L., He, G., Yang, M., Geng, S., Sun, Y., Jiang, S. (2025). Chitosan-loaded coconut fiber for highly-enhanced heavy metal adsorption from wastewater. *Separation and Purification Technology, 368*, 133105. https://doi.org/https://doi.org/10.1016/j.seppur.2025.133105.

Ye, L., Xu, Z., Li, Y., Wang, P., Yan, Y., & Yan, J. (2025). A chitosan-integrated antibacterial protein composite nanocomplex derived from barnacle cement and spider silk. *Journal of Biotechnology, 405*, 57-71. https://doi.org/https://doi.org/10.1016/j.jbiotec.2025.05.007.

Yıldırım, M., Poyraz, S., Acet, Ö., Acet, B. Ö., Karakoç, V., & Odabaşı, M. (2025). Chitosan hydrogels: Versatile platforms for drug delivery in cancer treatment, wound dressing, and 3D bioprinting applications. *International Journal of Biological Macromolecules, 314*, 144367. https://doi.org/https://doi.org/10.1016/j.ijbiomac.2025.144367.

Yilmaz, O. E., & Yilmaz, A. C. (2026). Green-fabricated enzyme-functionalized chitosan/Cu-BTC nanofibers for ultralow-level, room-temperature CO detection with enhanced selectivity and stability. *Talanta, 296*, 128509. https://doi.org/https://doi.org/10.1016/j.talanta.2025.128509.

Yonpiam, M., Wannigama, D. L., Shein, A. M. S., Liao, T., Hurst, C., Monk, P. N., Chatsuwan, T. (2025). Combination of curcumin or chitosan with photodynamic therapy as an effective alternative therapy for overcoming wound infection associated with multidrug-resistant Acinetobacter baumannii. *The Microbe, 7*, 100375. https://doi.org/https://doi.org/10.1016/j.microb.2025.100375.

Zanotti, A., Baldino, L., Iervolino, G., Cardea, S., & Reverchon, E. (2025). Chitosan Aerogel Beads Production by Supercritical Gel Drying, and Their Application to Methylene Blue Adsorption. *The Journal of Supercritical Fluids*, 106731. https://doi.org/https://doi.org/10.1016/j.supflu.2025.106731.

Zeng, H., Wu, M., Gao, Z., Ma, Y., Zhou, H., & Sun, J. (2025). Cross-linking oxidized rice straw and chitosan for preparing a fully bio-based adhesive with exceptional water-resistant bonding performance. *International Journal of Biological Macromolecules, 315*, 144416. https://doi.org/https://doi.org/10.1016/j.ijbiomac.2025.144416.

Zhang, H., Zhang, X., Wang, B., Ren, B., & Yang, X. (2025). Synergistic multi-crosslinked networks in chitosan/Gallic acid/Fe3+ composite biofilms with high UV resistance and antioxidant properties. *International Journal of Biological Macromolecules, 316*, 144686. https://doi.org/https://doi.org/10.1016/j.ijbiomac.2025.144686.

Zhang, J., Yang, J., Zhou, R., Liu, H., Chang, Y., & Babichuk, I. S. (2025). A polymer gel-gated flexible transistor based on chitosan physically cross-linked films for hydrogen sulfide gas sensing. *Sensors and Actuators A: Physical, 390*, 116593. https://doi.org/https://doi.org/10.1016/j.sna.2025.116593.

Zhang, Q., Wang, H., Chen, M., Wang, L., & Che, X. (2025). Color-changeable and separable PLA/PVP/chitosan tri-layer self-pumping dressing containing astragalus for promoting the healing of infectious burn wound. *International Journal of Biological Macromolecules, 321*, 146043. https://doi.org/https://doi.org/10.1016/j.ijbiomac.2025.146043.

Zhang, X., Li, Z., Pizzi, A., Qin, Z., Zhu, G., Li, X., Deng, S. (2025). Development of an environmentally friendly lignosulfonate-chitosan adhesive with excellent water resistance through epoxidation modification. *International Journal of Biological Macromolecules, 320*, 145846. https://doi.org/https://doi.org/10.1016/j.ijbiomac.2025.145846.

Zhang, X., Wu, S., Liu, R., Jiang, H., Jing, S., & Zhuang, H. (2025). Smart-responsive chitosan-based materials for precise degradation control and environmental remediation: A review. *International Journal of Biological Macromolecules, 319*, 144982. https://doi.org/https://doi.org/10.1016/j.ijbiomac.2025.144982.

Zhang, Y., Guo, H., Fu, C., Fang, Z., Xiong, B., Li, W., & Zhu, B. L. L. (2025). Chitosan nanoreactor modified with PNIPAAm as efficient catalyst for preparation of chiral boron compounds in aqueous phase. *International Journal of Biological Macromolecules, 311*, 143865. https://doi.org/https://doi.org/10.1016/j.ijbiomac.2025.143865.

Zhang, Z., Zhao, J., Li, W., Yuan, H., Chi, Y., Tang, J., Xie, Z. (2025). Efficient removal of anionic dyes by chitosan-based hydrogels modified with SDS and NaOH: Structural optimization, adsorption performance, and mechanistic insights. *International Journal of Biological Macromolecules, 321*, 146275. https://doi.org/https://doi.org/10.1016/j.ijbiomac.2025.146275.

Zhao, S., Wu, J., Li, M., Zhao, Y., Xiong, G., Wang, X., & Peng, L. (2025). Assessment of chitosan nanoemulsion loading with perillaldehyde on the quality characteristics and microbial diversity of beef slices during refrigeration storage. *Food Chemistry: X, 29*, 102658. https://doi.org/https://doi.org/10.1016/j.fochx.2025.102658.

Zheng, C., Sun, K., Wu, Q., Sun, Y., Xu, B., & Sun, L. (2025). Green synthesis of robust chitosan-based hydrogel beads for Cr(VI) removal from water. *Separation and Purification Technology, 377*, 134271. https://doi.org/https://doi.org/10.1016/j.seppur.2025.134271.

Zhou, D., Li, X., Zhu, Y., Hu, Y., Zhang, S., Tong, Z., Chen, Y. (2025). Fungal-derived chitosan-based hydrogels with antimicrobial properties for infectious wound healing. *Carbohydrate Polymers, 366*, 123917. https://doi.org/https://doi.org/10.1016/j.carbpol.2025.123917.

Zhou, R., Zhang, X., Wen, Y., Huo, W., Mu, Y., Liu, H., Li, J. (2025). One-step purification using MWCNTs@chitosan composite adsorbents for effective matrix interference removal in determination of 49 Pesticides in aquatic products. *Journal of Hazardous Materials, 496*, 139295. https://doi.org/https://doi.org/10.1016/j.jhazmat.2025.139295.

Zhou, X., Wu, S., Liu, P., Wang, L., & Xie, F. (2025). Hybrid bioink of methyacrylated starch with minimal methacrylated chitosan enables high-precision 3D printing for complex tissue scaffolds. *Carbohydrate Polymers, 367*, 124023. https://doi.org/https://doi.org/10.1016/j.carbpol.2025.124023.

Zhou, Y.-Q., Deng, X., Zhao, Z.-X., Wang, X.-T., Jin, X.-Q., Xiong, J.-B., Yin, D.-C. (2025). High-strength chitosan-sericin cryogel with synergistically reinforced networks for hemostasis. *International Journal of Biological Macromolecules, 318*, 145038. https://doi.org/https://doi.org/10.1016/j.ijbiomac.2025.145038.

Zhou, Y., Zhang, Y., Nie, Y., Sun, D., Wu, D., Ban, L., Pan, X. (2025). Recent advances and perspectives in functional chitosan-based composites for environmental remediation, energy, and biomedical applications. *Progress in Materials Science, 152*, 101460. https://doi.org/https://doi.org/10.1016/j.pmatsci.2025.101460.

Zhu, H., Xu, Z., Xiao, M., Zhao, D., Wang, Q., Fu, Y., Jiang, R. (2025). Multifunctional and sustainable chitosan-based interfacial materials for effective water evaporation, desalination, and wastewater purification: A review. *International Journal of Biological Macromolecules, 321*, 145973. https://doi.org/https://doi.org/10.1016/j.ijbiomac.2025.145973.

Zhuravleva, I. L., Berezin, B. B., Piskarev, V. E., Bezrodnykh, E. A., Kulikov, S. N., Tikhonov, V. E., & Antonov, Y. A. (2026). Interaction of phosvitin with Chitosan: Effect of pH, ionic strength, and temperature. *Food Hydrocolloids, 171*, 111751. https://doi.org/https://doi.org/10.1016/j.foodhyd.2025.111751.

Chapter 2

Three (D)imensions of Chitosan: Applications in Dialysis, Dermatology and Diabetes

Maria Martingo[1,*]
Maria Leonor de Castro[1]
Ana Rita Pereira[2]
Manuela Pintado[1]
Patrícia Henriques[3]
Sara Baptista-Silva[1]
and Sandra Borges[1]
[1]Católica Portuguesa, CBQF - Centro de Biotecnologia e Química Fina – Laboratório Associado, Escola Superior de Biotecnologia, Porto, Portugal
[2]BioAscent Discovery Ltd, Bo'ness Road, Newhouse, Scotland
[3]GOTECH Antimicrobial, Porto, Portugal

Abstract

Chitosan, a linear polysaccharide composed of β-(1→4)-linked D-glucosamine and N-acetyl-D-glucosamine units, is a biopolymer valued for its biocompatibility, biodegradability, broad-spectrum antimicrobial, hemostatic, and anti-inflammatory properties, and huge versatility in biomedical formulations. It can be obtained by both green and conventional methods, offering broad potential across biomedical, dermatological, and metabolic applications.
In the biomedical arena and regarding hemodialysis treatment, chitosan-based membranes have shown promising potential due to their

* Corresponding Author's Email: marialimamartingo@gmail.com

In: Chitosan
Editor: Denis Hurst
ISBN: 979-8-90134-079-0

biocompatibility and selective permeability. The tunable structure improved toxin removal, multifunctionality and sustainability compared to conventional dialysis materials (e.g., polysulfone, polyethersulfone). This adaptability illustrates how structural tailoring translates into tangible clinical benefits. The same multifunctionality has been extensively explored in dermatology, where chitosan properties have been harnessed for the treatment of wounds and skin diseases, demonstrating a keen interest in the design of cutting-edge therapeutic solutions. Moreover, its nature makes it an excellent matrix for barrier creams and transdermal drug-delivery systems that have the ability to protect the skin and release active agents in a controlled way. Beyond localized therapies, chitosan also shows systemic relevance in metabolic health. Recent studies have highlighted chitosan's ability to protect and promote the proliferation of pancreatic β-cells, reduce hyperglycemia, and prevent lipid metabolism disorders commonly associated with diabetes mellitus.

This chapter aims to present a comprehensive overview of recent developments and potential biomedical applications of chitosan, based on an in-depth analysis of current research and technological advancements. Future research should focus on advanced functionalization, optimization of green production methods, and translational studies to fully harness their therapeutic benefits.

Keywords: chitosan multifunctionality, chronic renal disease, skin conditions, metabolic diseases

1. Introduction

Chitosan has emerged as one of the most versatile natural biopolymers for biomedical innovation. It is obtained through the partial deacetylation of chitin and is composed of β-(1→4)-linked D-glucosamine and N-acetyl-D-glucosamine units, distinguished by the presence of primary amino groups at the C-2 position that become protonated under mildly acidic conditions (K. Li et al., 2016; Muxika et al., 2017). This feature, together with the ability to tailor the degree of deacetylation and molecular weight, governs solubility, charge density, reactivity, and ultimately the polymer's performance in biological environments. The degree of deacetylation and molecular weight determine not only solubility and viscosity but also biological activity, with high molecular weight forms conferring structural robustness and low molecular weight derivatives offering superior solubility and bioavailability.

The chemical architecture of chitosan, with reactive amino and hydroxyl groups, further allows for modifications such as quaternization, sulfonation, grafting, or nanoparticle incorporation, which extend its applications beyond the limitations of the native material (Aranaz et al., 2021; Muxika et al., 2017; Sogias et al., 2010a).

The biomedical relevance of chitosan is evident in the direct link between its chemistry and biological performance. To contextualize its potential, this chapter follows what can be called a "three dimensions" approach, focusing on dialysis, dermatology, and diabetes.

In dialysis, protonated amino groups facilitate the adsorption of negatively charged uremic toxins and phosphates, enabling chitosan to complement diffusion and convection with an adsorptive mechanism in blood purification. Chitosan addresses the shortcomings of conventional synthetic membranes by adding adsorption capacity for protein-bound toxins, reducing platelet adhesion, and improving hemocompatibility when used as a coating, blend, or grafted derivative (Dai et al., 2011; Ji et al., 2023; Pereira & Bártolo, 2014; Radu & Voicu, 2022; Zawada et al., 2022).

In dermatology, its antimicrobial activity and hemostatic capacity support infection control and wound healing, while its mucoadhesive properties enhance non-invasive drug delivery by increasing paracellular permeability. Its antimicrobial, anti-inflammatory, film-forming, and regenerative properties are harnessed in nanoparticles, nanofibers, and hydrogels for the treatment of psoriasis, atopic dermatitis, seborrheic dermatitis, acne, chronic wounds, and even skin cancer (Herdiana et al., 2023; Victor et al., 2020).

In diabetes, the conversion of chitosan into chitooligosaccharides further improves solubility and bioavailability, with reported antioxidant, anti-inflammatory, and protective effects on pancreatic β-cells, as well as improvements in glucose handling through modulation of hepatic and muscular pathways. While protecting β-cells from oxidative stress, chitooligosaccharides improve insulin sensitivity, and contribute to better glucose homeostasis, also serving as carriers for antidiabetic drugs and inhibitors of lipid absorption (Pandya et al., 2016; Wang et al., 2024).

In the pursuit of more sustainable and biocompatible production routes, alternative sources and innovative processing strategies for chitosan have gained increasing attention. Fungal-derived chitosan and green extraction methods such as enzymatic deproteinization, fermentation, or the use of ionic liquids provide sustainable and allergen-free alternatives, reinforcing its alignment with both biomedical and environmental priorities (Abasian et al., 2020; Davis et al., 2024; Huq et al., 2022).

Through this perspective, chitosan is not seen as a single universal solution but as a platform material whose properties can be engineered to meet specific clinical needs. Its molecular structure can be fine-tuned, its functionality extended through chemical modification, and its processing adapted to create devices and formulations that address pressing medical challenges. By framing its development within the three dimensions of dialysis, dermatology, and diabetes, this chapter provides a coherent overview of how chitosan's molecular promise can be translated into tangible biomedical impact.

2. Chitosan: Structure, Properties and Extraction Methods

Chitosan is a linear polysaccharide obtained through the partial deacetylation of chitin, a natural biopolymer abundantly present in the exoskeletons of crustaceans, insect cuticles, and fungal cell walls (Hafsa et al., 2021; Nasr, Rady, Gomaa, Syrovet, et al., 2019). Chitosan stands out as the only naturally occurring cationic polysaccharide, a property that underpins many of its physicochemical and biological activities (K. Li et al., 2016; Muxika et al., 2017).

Two molecular parameters largely govern the behavior of chitosan: the degree of deacetylation and the molecular weight. The degree of deacetylation, usually ranging between 50% and 95%, determines the proportion of free amino groups along the chain and therefore modulates solubility, charge density, and chemical reactivity. A higher degree of deacetylation generally enhances solubility in acidic environments and increases electrostatic interactions with negatively charged molecules such as proteins, nucleic acids, and uremic toxins (Nguyen et al., 2022). The molecular weight, which spans from a few kDa to over 1,000 kDa, affects viscosity, mechanical strength, and biological activity. While high-molecular-weight chitosan confers strong film-forming ability and structural stability, low-molecular-weight forms are preferred in biomedical contexts for their greater solubility, lower viscosity, and enhanced biocompatibility (Alemu et al., 2023; Nguyen et al., 2022; Suryani et al., 2022). Furthermore, the reactive hydroxyl groups at C-3 and C-6, combined with the amino group at C-2, provide multiple sites for derivatization. Through chemical modifications such as quaternization (Marin et al., 2023; Yan & Qiu, 2024), sulfonation, acylation, or graft copolymerization (X. Feng et al., 2024), the functional scope of chitosan can be extended beyond its native limitations of pH-dependent solubility and moderate chemical stability.

The molecular structure of chitosan directly translates into a versatile physicochemical profile, with high biodegradability through enzymatic action by chitosanases and lysozyme yields non-toxic oligomers and glucosamine, which can be resorbed by biological systems. Thermal stability is moderate, with decomposition occurring above 250°C, but this threshold is influenced by the degree of deacetylation and the molecular weight (Aranaz et al., 2021; Muxika et al., 2017; Sogias et al., 2010).

Equally important are its biological properties, which make chitosan particularly attractive in medical and pharmaceutical applications. Chitosan is broadly recognized for its biocompatibility and low cytotoxicity. The protonated amino groups confer antimicrobial activity, as they disrupt negatively charged bacterial membranes, alter nutrient uptake, and interact with microbial DNA (Gheorghiță et al., 2023; Gonciarz et al., 2025; Kucharska et al., 2019; Rabea et al., 2003; Sogias et al., 2010). In addition, chitosan demonstrates hemostatic activity by enhancing platelet aggregation and accelerating coagulation, explaining its frequent use in wound dressings and surgical applications (Bowry et al., 2021). Its anti-inflammatory and immunomodulatory roles have also been reported, with evidence that it reduces oxidative stress and cytokine production *in vitro* and *in vivo*. Furthermore, chitosan's mucoadhesive properties enable it to transiently increase paracellular permeability across epithelial barriers, a feature that has been widely exploited for mucosal drug delivery systems (Bhattarai et al., 2010; Patel et al., 2010). Together, these physicochemical and biological properties derive directly from its molecular architecture, highlighting the interplay between structure and functionality and reinforcing its potential as a multifunctional biomaterial.

The practical availability of chitosan is determined not only by its natural abundance but also by the extraction and processing methods employed. The most common sources of chitin are crustacean shells, particularly shrimp, crab, and lobster, although insects and fungi also provide alternative supplies (Khanafari et al., 2008). Conventional industrial production relies on a sequence of three chemical steps: demineralization, deproteinization, and deacetylation. Demineralization typically involves treatment with hydrochloric acid to remove calcium carbonate and other minerals, while deproteinization relies on sodium hydroxide to eliminate proteins and residual organic matter (Chatterjee et al., 2005; Davis et al., 2024). The resulting chitin is then subjected to alkaline deacetylation, commonly with concentrated sodium hydroxide at elevated temperatures, to convert acetyl groups into free amino groups and thereby produce chitosan. This method is efficient and scalable, but it consumes large amounts of harsh chemicals, generates considerable waste, and can degrade polymer chains, lower molecular weight and altering functionality.

The environmental and structural limitations of conventional methods have prompted the development of alternative green and enzymatic approaches (Ahmed et al., 2022). Enzymatic deproteinization employs proteolytic enzymes such as alcalase, papain, or pepsin to replace harsh alkaline treatments, improving selectivity and reducing protein contamination. Similarly, deacetylation can be achieved using chitin deacetylase, though enzymatic processes are often slower and less efficient at an industrial scale. Other green strategies include fermentation-based extraction, in which lactic acid bacteria produce organic acids that simultaneously demineralize and deproteinize crustacean shells, yielding high-quality chitin with minimal chemical inputs (R. Castro et al., 2018; Gortari & Hours, 2013; Khanafari et al., 2008). Ionic liquids and deep eutectic solvents have also been investigated as environmentally benign solvents for chitin extraction and deacetylation, offering milder processing conditions and potential recyclability of solvents. In addition, fungal chitosan derived from the cell walls of species, such as *Aspergillus niger* or *Mucor rouxii,* provides a sustainable alternative, eliminating reliance on marine crustaceans and reducing allergenicity concerns. Although these green and enzymatic methods currently face challenges in scalability and cost, they represent promising directions for the sustainable and controlled production of chitosan with tailored molecular characteristics (Abasian et al., 2020; R. Castro et al., 2018; Chatterjee et al., 2005; Davis et al., 2024; Gortari & Hours, 2013; Khanafari et al., 2008).

3. Chitosan in Dialysis

3.1. Chronic Renal Disease: Clinical Needs and Hemodialysis Requirements

The kidneys are the body's excretory system, which eliminates toxins, excess fluid, and waste products from metabolism. A variety of related health issues can arise from impaired kidney function, which causes an excess of fluids and waste products (Martingo, Baptista-Silva, Pintado, et al., 2025).

Chronic kidney disease (CKD) represents a major and rapidly growing global public health challenge, currently affecting more than 10% of the world's population, and is projected to become the 5th leading cause of death by 2040 (Martingo, Baptista-Silva, Mesquita, et al., 2025).

As renal function declines, the kidneys progressively lose their ability to excrete metabolic waste products, regulate electrolyte balance, and maintain acid–base homeostasis. In advanced stages (CKD stage 5) (Dhondup & Qian, 2017), patient survival depends on renal replacement therapies, namely hemodialysis (HD), peritoneal dialysis (PD), or kidney transplantation (Muñoz-Terol et al., 2023). Among these, HD is the most widely adopted modality, serving as the mainstay of treatment for millions of patients worldwide (Teuwafeu et al., 2024).

Renal failure leads to the systemic accumulation of uremic toxins, which include small molecules such as urea and creatinine, middle molecules such as β2-microglobulin, and protein-bound solutes including p-cresyl sulfate and indoxyl sulfate. Their accumulation is directly associated with uremic symptoms, heightened cardiovascular risk, and chronic inflammatory processes that contribute to morbidity and mortality (Shafiee et al., 2024; Yamamoto, 2019). Importantly, long-term survival in HD patients remains markedly reduced, with annual mortality rates of 15–20%, surpassing those of many common cancers (Msaad et al., 2019).

HD only partially addresses these needs by circulating blood extracorporeally through a semipermeable membrane in contact with a dialysate solution (Engelen et al., 2023). The removal of solutes relies on three key mechanisms: diffusion, which efficiently clears small molecules; convection, which supports the transport of middle molecules (Macías et al., 2019); and adsorption, increasingly recognized as critical for hydrophobic and protein-bound compounds (Morisi et al., 2025).

Clinical complexity dictates the essential requirements of modern dialysis membranes. These include high biocompatibility to minimize complement activation, coagulation, and inflammation (Ávila et al., 2025); selective permeability to enable efficient clearance of small and middle molecules while preserving vital proteins such as albumin; adsorption capacity to target protein-bound toxins; and robust mechanical and chemical stability to withstand long treatment cycles. More recently, the sustainability of dialysis technologies has emerged as an additional challenge, given the significant ecological footprint of HD, which consumes large amounts of ultrapure water and generates substantial plastic waste (Backston et al., 2025).

The relationship between clinical needs in CKD and the corresponding requirements for HD membranes is summarized in Table 1.

Table 1. Clinical needs in chronic kidney disease and their implications for hemodialysis membrane design

Clinical need	Clinical example	Implication for HD membranes	References
Accumulation of small solutes	Urea, creatinine	High diffusivity	(Pandya et al., 2016)
Retention of middle molecules	β2-microglobulin	Controlled permeability via convection	(Wolley et al., 2018)
Presence of protein-bound toxins	Indoxyl sulfate, p-cresyl sulfate	Effective adsorption capacity	(Vanholder et al., 2014)
Chronic inflammation induced by blood–material contact	Complement activation, pro-inflammatory cytokines	Optimized biocompatibility	(Salehi Moghaddam et al., 2025)
Risk of loss of essential proteins	Serum albumin	Rigorous selectivity	(Shannon et al., 2021)
High cardiovascular mortality	Phosphate retention, oxidative stress	Broader uremic toxin clearance	(Yang et al., 2023)
Environmental impact of HD	High water consumption, plastic waste	Development of eco-friendly, sustainable membranes	(Martingo, Baptista-Silva, Pintado, et al., 2025)

While synthetic membranes represent the current benchmark in HD, their environmental footprint highlights the need for alternative solutions (Arias-Guillén et al., 2024). In this context, the exploration of chitosan-based biomaterials may open new opportunities to design membranes that not only meet, but potentially exceed, clinical requirements, combining efficacy, safety, and sustainability.

3.2. Conventional Dialysis Membranes: Limitations

Since the introduction of dialysis therapy in the mid-20th century, membranes have been central to its effectiveness and safety (Blagg, 2007). Early devices employed regenerated cellulose films such as cuprophan, which enabled the clinical establishment of HD, but were limited by poor biocompatibility, frequent complement activation, and induction of inflammatory reactions (Eduok et al., 2021). Advances in polymer chemistry led to the development of synthetic membranes, including polysulfone (PSu), polyethersulfone (PES) and polyacrylonitrile (PAN), which today dominate clinical practice, however, with some critical limitations. These conventional membranes efficiently eliminate small solutes such as urea and creatinine, nonetheless, for some larger middle-molecule toxins (e.g., β2-microglobulin), the clearance efficiency of the conventional high-flux membranes remains limited. Therefore, for the effective clearance of middle-molecule toxins, more advanced blood purification technologies, such as high-cutoff dialysis or adsorption-based methods (e.g., hemoadsorption), are often required (Cozzolino et al., 2025). Moreover, the reduction of β2-microglobulin, although improved with high-flux technologies, has not eliminated dialysis-related amyloidosis, a long-term complication in chronic patients (Scarpioni et al., 2016).

Biocompatibility issues, although reduced, are not fully resolved. Contact between blood and synthetic polymers can still activate the complement system, coagulation cascades, and leukocytes, contributing to chronic inflammation and oxidative stress in long-term patients (Skinner et al., 2021). Additionally, the adsorption capacity of most synthetic membranes is limited, restricting their ability to capture hydrophobic or protein-bound solutes (Vega-Vega et al., 2023). Attempts to modify synthetic polymers, for example by coating PSu membranes with vitamin E to reduce oxidative stress or modifying PAN with polyethyleneimine to enhance endotoxin adsorption (J. Zhang et al., 2025), have shown some promise but remain only partial solutions, without fully addressing the spectrum of uremic toxins or the systemic burden of inflammation (Mydlík et al., 2004; H. Zhang et al., 2014).

Another major limitation lies in the balance between permeability and selectivity. Increasing pore size and membrane permeability enhances the removal of middle molecules but simultaneously increases the risk of losing vital proteins such as albumin. Hypoalbuminemia, a frequent condition in dialysis patients, is a strong predictor of mortality and hospitalization (Soeters et al., 2019). Therefore, clinicians and engineers face a trade-off: maximizing clearance of toxic solutes without compromising essential plasma proteins. Conventional synthetic membranes have not yet achieved the level of selectivity that mimics the glomerular filtration barrier of the natural kidney (Kalantar-Zadeh et al., 2021).

Finally, there is increasing recognition of the environmental burden associated with conventional membranes. HD requires vast amounts of ultrapure water, often between 300 and 500 liters per session, and produces significant volumes of plastic waste, much of it derived from petroleum-based polymers (Coulliette & Arduino, 2013). With millions of patients worldwide undergoing dialysis three times per week, the ecological footprint of current dialysis technologies is substantial. Sustainability considerations, previously overlooked, are now becoming an integral part of the discussion on the future of renal replacement therapies (Laranjinha et al., 2023).

Conventional dialysis membranes have been essential in transforming CKD management and prolonging survival, but the ecological concerns linked to their large-scale use collectively highlight the need for innovative solutions. Natural-derived biomaterials such as chitosan, with their favorable bioactivity, selective binding capacity, and potential for sustainable production, may represent a paradigm shift in the design of HD membranes.

3.3. Chitosan-Based Membranes in Hemodialysis

The limitations of conventional dialysis membranes have prompted interest in natural and functional biomaterials. Among these, chitosan has attracted increasing attention as a versatile candidate for dialysis membranes (Desai et al., 2023). When obtained from fungal fermentation, chitosan is free from animal-derived allergens and produced under more sustainable conditions, aligning with both biomedical and environmental requirements (Huq et al., 2022). Its intrinsic characteristics provide a rationale for its use in blood-contacting devices such as HD membranes, including antimicrobial activity (Ke et al., 2021), hemostatic effects (Gheorghiță et al., 2023), biodegradability, and the ability to interact electrostatically with negatively charged molecules (Baroudi et al., 2023; Mirda et al., 2021; Monteiro & Airoldi, 1999; Wu et al., 1999). These features suggest that natural-derived polymers could complement the performance of existing synthetic membranes. A comparative overview is provided in Table 2.

3.3.1. Adsorption of Uremic Toxins

One of the most relevant properties of chitosan is its capacity for adsorption. The positively charged amino groups of chitosan interact with negatively charged solutes, including phosphate (Viala et al., 1996), uric acid (Noureen et al., 2025), and protein-bound uremic toxins such as indoxyl sulfate (Anraku et al., 2012; Hsu et al., 2024; Sirisuksakun et al., 2025) and p-cresyl sulfate (Xiong et al., 2021). This property directly addresses one of the central shortcomings of conventional synthetic membranes, which exhibit limited adsorption capacity. Chitosan-based membranes can therefore complement diffusion and convection mechanisms with an additional adsorption pathway, broadening the spectrum of toxin removal. In addition, chitosan has demonstrated the ability to chelate metal ions (e.g., iron, copper, zinc) and to bind organic compounds through hydrogen bonding and hydrophobic interactions, further enhancing its (Verma & Quraishi, 2021; Wu et al., 1999). Studies have shown that chitosan-functionalized membranes or composites can capture middle molecules and toxins more effectively than unmodified PSu membranes, suggesting a synergistic effect when chitosan is incorporated as a surface coating or a blended polymer (Baroudi et al., 2023; Bhatt et al., 2023; Dambuza et al., 2025).

Table 2. Comparison of conventional and emerging hemodialysis membranes

Membrane type	Examples	Advantages	Limitations	Current status	References
Regenerated cellulose	Cuprophan®, cellulose triacetate	Widely available, cost-effective, early standard in HD	Poor biocompatibility, complement activation, limited toxin clearance	Non used	(Ávila et al., 2025)
Synthetic polymers	polysulfone (PSu), polyethersulfone (PES), polyacrylonitrile (PAN)	High permeability, improved biocompatibility, high-flux options for middle molecules	Potential immune activation, environmental burden	Current clinical standard worldwide	(Arias-Guillén et al., 2024; Kobayashi & Ohtake, 2023; Mukherjee et al., 2008; Tarrass & Benjelloun, 2022)
Modified synthetic membranes	PAN with polyethyleneimine coating, vitamin E-coated PSu, chitosan-blended	Improved adsorption, antioxidant properties, reduced oxidative stress	Increased cost, partial improvements only	In research	(Lin et al., 2024; Martingo, Baptista-Silva, Mesquita, et al., 2025; Martingo, Baptista-Silva, Pintado, et al., 2025; Mydlík et al., 2004; Shi & Chang, 2022; H. Zhang et al., 2014)

3.3.2. Biocompatibility and Anti-Inflammatory Effects

Biocompatibility remains a major challenge in dialysis membranes, as repeated blood–material contact induces complement activation, coagulation, and leukocyte stimulation. Chitosan's biocompatible and bioactive profile provides an advantage in this regard. Experimental studies indicate that chitosan reduces platelet adhesion and can attenuate activation of inflammatory markers when compared to conventional synthetic polymers (Bhattarai et al., 2010). Its natural hemostatic capacity, widely used in wound dressings and surgical applications, reflects a controlled interaction with coagulation pathways rather than uncontrolled activation (Chou et al., 2003; Gheorghiță et al., 2023; Lord et al., 2011). Furthermore, in their recent work, (Valades-Aguilar et al., 2024) developed Psu membranes grafted with sulfonated dihydroxypropyl chitosan (SDHPCS) using UV-induced grafting as a strategy to enhance hemocompatibility. Surface analyses confirmed the successful incorporation of the sulfonated chitosan derivative, which significantly increased membrane hydrophilicity and resulted in smoother surface morphology. When tested with human plasma, the modified membranes exhibited markedly reduced protein adsorption and platelet adhesion, accompanied by a clear decrease in fibrin-related deposition compared with pristine Psu membranes. These properties not only support its use as a structural membrane material but also as a functional additive to improve the hemocompatibility of synthetic membranes (Chou et al., 2003; Gheorghiță et al., 2023; Lord et al., 2011; Moran et al., 2018; Valades-Aguilar et al., 2024).

3.3.3. Functional Modifications and Composite Membranes

While native chitosan possesses a range of advantageous properties, including biocompatibility, biodegradability, and intrinsic bioactivity, its direct application as a dialysis membrane is limited by certain shortcomings. Its relatively low mechanical strength, restricted solubility at physiological pH, and insufficient long-term stability under blood-contacting conditions represent significant challenges. To address these limitations, a variety of modification strategies have been investigated (Table 3). Crosslinking chitosan with agents such as glutaraldehyde, genipin, or tripolyphosphate has been shown to enhance its structural integrity and resistance to dissolution, thereby improving performance under aqueous and physiological environments (Du et al., 2020; Hu et al., 2024; Monteiro & Airoldi, 1999). Another widely explored approach involves grafting or blending chitosan with synthetic polymers such as PSu and PES, which imparts the necessary mechanical robustness while preserving the desirable bioactivity of chitosan (Yan & Qiu, 2024). Studies on surface modifications, including sulfonation (Yan & Qiu, 2024) and quaternization (Marin et al., 2023), have shown that chitosan nanofibers display improved solubility, antimicrobial activity, and hemocompatibility, thereby emerging as promising materials for HD membranes (Marin et al., 2023). Nanocomposite strategies that incorporate chitosan with nanoparticles such as silver, titanium dioxide, or graphene oxide have expanded the functional landscape of these membranes, providing additional antimicrobial protection, antioxidant properties, and improved adsorption selectivity (W. Feng & Wang, n.d.; Mesgari et al., 2021; Mirda et al., 2021). Collectively, these modifications have enabled the development of hybrid membranes that merge the durability of conventional synthetic polymers with the multifunctional bioactivity of chitosan, positioning them as a promising avenue for next generation of HD devices.

Table 3. Strategies for modifying chitosan to improve its performance in HD membranes

Strategy	Mechanism	Advantages	Limitations / Challenges	References
Crosslinking (e.g., glutaraldehyde, genipin, tripolyphosphate)	Formation of covalent or ionic crosslinks between chitosan chains	Improves mechanical strength, stability, and resistance to dissolution in aqueous/physiological media	Potential cytotoxicity may reduce flexibility and bioactivity	(Du et al., 2020; Hu et al., 2024; Monteiro & Airoldi, 1999)
Blending / grafting with synthetic polymers (e.g., PSu, PES)	Physical mixing or chemical grafting of chitosan with robust synthetic matrices	Combines durability of synthetics with chitosan's biocompatibility; scalable for industrial processing	Possible phase separation; dilution of chitosan's surface activity	(Yan & Qiu, 2024)
Surface modifications (e.g., sulfonation, quaternization)	Chemical substitution of amino/hydroxyl groups	Enhances hydrophilicity; improves selective adsorption of negatively charged uremic toxins	Multi-step synthesis; batch variability; cost	(Marin et al., 2023; Yan & Qiu, 2024)
Nanocomposites (e.g., chitosan + Ag, TiO_2, graphene oxide)	Incorporation of nanoparticles within chitosan matrix	Adds antimicrobial, antioxidant, and adsorption properties; potential for multifunctional membranes	Nanoparticle leaching/toxicity; regulatory and safety concerns; reproducibility	(W. Feng & Wang, n.d.; Mesgari et al., 2021; Mirda et al., 2021)

3.4. *In Vitro* and *In Vivo* Studies of Chitosan-Based Membranes in Hemodialysis

Research on chitosan-based membranes has largely focused on *in vitro* models, but an increasing number of studies also explore their performance *in vivo*. *In vitro* assays consistently demonstrate enhanced adsorption of uremic toxins, improved hemocompatibility, and antimicrobial effects, which are beneficial in reducing infection risks associated with dialysis equipment (Almeida et al., 2025; Lin et al., 2024; Yadav et al., 2024). For example, *in vitro* hemoperfusion experiments using chitosan-based adsorbents showed improved clearance of indoxyl sulfate and β2-microglobulin compared to PSu controls. Similarly, recent advances showed that PSu membranes grafted with sulfonated chitosan showed reduced protein and platelet adhesion together with prolonged coagulation times, reflecting improved hemocompatibility in plasma assays (Yan & Qiu, 2024).

Recent reviews of animal models of kidney disease emphasize both their indispensability and their limitations, underscoring the need to complement rodent studies with alternative models and human-derived systems to better capture the complexity of CKD and improve translational relevance (Liang & Liu, 2023). Despite these encouraging results, translation to routine clinical use remains limited. Issues such as large-scale reproducibility, regulatory approval, and long-term safety must still be resolved.

4. Chitosan in Dermatology and Skin Health

4.1. Skin Conditions

The skin is the largest organ of the human body (including the hypodermis), covering approximately 2 m^2 and accounting for up to 16% of the total body weight (~3.6 kg in adults, twice the brain's weight). It consists of two main layers – the epidermis and dermis – and includes appendages such as glands, hair follicles, and nails. Acting as the body's primary interface with the environment, the skin provides sensory perception, protection, and homeostasis by regulating temperature, fluids, vitamin D, fat storage, blood pressure, and excretion (M. L. Castro et al., 2023). Its integrity can be compromised by several external and internal aggressions, including inflammatory diseases, such as psoriasis, atopic dermatitis, seborrheic dermatitis and acne vulgaris, wounds, and cancer. In general, inflammatory diseases can be described as a range of immune-mediated conditions, each with specific clinical and histopathological features. They are regarded as complex genetic diseases, arising from interactions between multiple genes as well as between genetic predisposition and environmental factors (Bieber, 2023). Wounds are defined as injuries or disorders that compromise the structural integrity of the skin, and are caused by external factors such as cuts, burns, or pressure, as well as from surgical interventions or underlying pathological conditions, including diabetes (Kolimi et al., 2022). Lastly, skin cancer is among the most common malignancies worldwide, with incidence and mortality rising, particularly in populations with white-skinned inhabitants (Roky et al., 2025).

4.2. Chitosan-Based Formulations for Skin Therapeutics

Research on chitosan and its application in skin health and treatments has been increasing exponentially over the last ten years, which reflects the growing interest in natural, sustainable and new biopolymers for cosmetic and cosmeceutical purposes, highlighting the scientific relevance of chitosan. Compared to conventional therapies, chitosan offers additional benefits, including biocompatibility, biodegradability, and the ability to promote skin regeneration, while also acting as an antimicrobial agent and an efficient vehicle for controlled drug delivery.

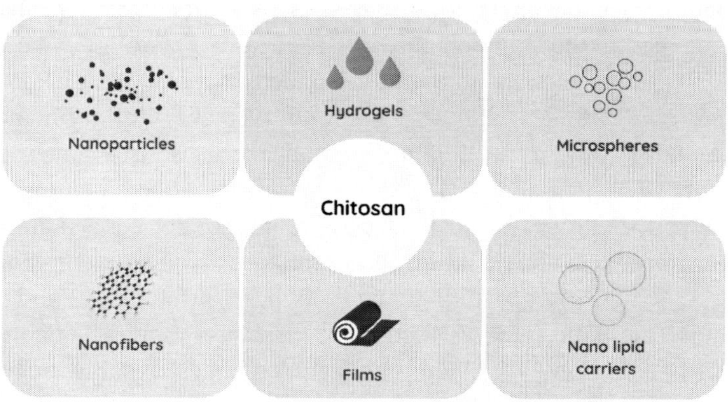

Figure 1. Examples of chitosan-based materials for the treatment of skin diseases.

Figure 1 presents some of the possible delivery systems incorporating chitosan that are being explored for the potential treatment of the skin pathologies mentioned.

4.2.1. Inflammatory Skin Conditions

Chitosan-based materials show great potential for the treatment of skin conditions. In psoriasis and atopic dermatitis, they improved lesion severity, lowered inflammation, and enhanced drug delivery in animal models. For seborrheic dermatitis, chitosan's antimicrobial, anti-inflammatory, and barrier-protective effects help restore skin balance. In acne vulgaris, nanocarriers reduced lesions and inflammation while maintaining high drug encapsulation and safety.

Psoriasis is a chronic, progressive and genetically influenced immune-mediated inflammatory disorder, presenting both cutaneous and systemic manifestations that may greatly impair patient's quality of life. Current topical treatments include corticosteroids, vitamin D analogues (e.g., calcitriol, calcipotriol), topical retinoids (e.g., tazarotene), calcineurin inhibitors (e.g., tacrolimus). Systemic therapies become necessary when topical approaches lose effectiveness, when lesions increase in extent or severity, or when symptoms cause significant discomfort. In such cases, treatment options involve drugs such as methotrexate, cyclosporine, and acitretin, phototherapy, or the use of certain biological agents (e.g., TNF inhibitors, IL-17 inhibitors) (Mascarenhas-Melo et al., 2022; Mohd Nordin et al., 2021; Zhu et al., 2022).

Concerning this condition, a study showed that tacrolimus-loaded chitosan nanoparticles, which allowed for an entrapment efficiency of 65.5%, concomitant with an enhanced local drug skin deposition *ex vivo*, with 82% of the drug retained in the skin compared to 34% from a tacrolimus® ointment. Moreover, using an *in vivo* imiquimod-induced psoriasis mouse model, the formulated nanoparticles allowed for improvements in erythema, skin thickness, and scaling, as well as in the total Psoriasis Area Severity Index (PASI) score (Fereig et al., 2021). Another study described a chitosan-based double-network hydrogel patch loaded with antioxidant ceria nanoparticles and betamethasone, which allowed significant reduction of epidermal hyperplasia, suppressed keratinocyte hyperproliferation, and lower inflammatory cytokine (Yeo & Kim, 2025) induced psoriatic mice (Yeo & Kim, 2025). Additionally, hyaluronic acid engineered gallic acid-loaded chitosan nanoparticles were reported to ameliorate psoriasis in an *in vivo* psoriasis mouse model, with reduced erythema, scaling, skin thickness, and total PASI score, presenting better results than a plain gallic acid formulation.

Also, the nanoparticle formulation proved to be efficient, by presenting an entrapment efficiency of 93.24% and a drug loading value of 73.17% (Sheikh et al., 2023).

Atopic dermatitis is the most common chronic inflammatory skin condition, and is marked by persistent itching, recurrent lesions and skin dryness. Often arising in childhood and sometimes continuing into adulthood, it shows diverse clinical patterns influenced by age, ethnicity and severity (Chuah et al., 2023; Prados-Carmona et al., 2025). Topical treatments include moisturizers, corticosteroids (e.g., methylprednisolone, prednisolone), and calcineurin inhibitors (e.g., pimecrolimus, tacrolimus), used alone or in combination. Other strategies, like wet wrap therapy and phototherapy are also considered. In contrast, systemic options are reserved for severe or treatment-resistant cases and involve non-biological agents, such as corticosteroids, cyclosporine, methotrexate, mycophenylate mofetil, and azathioprine, as well as of biological ones (Calabrese et al., 2022; Elahi et al., 2024; Perzynski, 2025).

In a topic dermatitis, chitosan/alginate-based nanoparticles loaded with tacrolimus were reported to significantly ameliorate related symptoms in 2,4-dinitrofluorobenzene-challenged rats, by reducing skin lesion score, ear thickness, IL-4 and IgE levels, and the number of mast cells in the treated group. *In vitro* studies corroborated these findings, as the nanoparticles allowed for the decrease of several inflammatory markers (IL-6, IL-8, TNF-α, and IFN-γ) in TNF-α/IFN-γ-challenged HaCaT cells. Further, the nanoparticles were able to facilitate drug penetration through the skin barrier, thereby enhancing drug accumulation in the skin, while presenting an encapsulation efficiency between 75 and 80% (Y. A. Huang & Lin, 2025). In another study, *in vitro* experiments revealed the anti-proliferative effect of tacrolimus-loaded chitosan-based nanoparticles over HaCaT cells. Also, using an *in vivo* house dust mites-induced atopic dermatitis mouse model, results revealed that topical administration of the nanoparticles allowed to alleviate symptoms such as ear thickness, epidermal/dermal thickness, IgE production, and severe inflammation, while reducing atopic dermatitis-related pro-inflammatory cytokines/chemokines in ear tissue, as well as downregulating the expression of Th2 response in ear tissue and serum (J. S. Lee et al., 2024). Furthermore, employing an *in vivo* house dust mite extract-induced atopic dermatitis mouse model, stem cell secretome-loaded nano-villi chitosan microspheres were shown to effectively reduce atopic dermatitis clinical severity compared to groups treated with plain secretome. Histological analysis revealed that the microspheres could suppress epidermal hyperplasia, immunocyte infiltration and mast cell activation (Oh et al., 2024).

In addition to atopic dermatitis, seborrheic dermatitis is a common and recurrent skin disorder marked by redness, flaking, itching, and an oily appearance. It mainly affects sebaceous gland-rich sites, such as the scalp, face, chest and back. Although not completely understood, its cause is multifactorial, involving sebum overproduction, *Malassezia* proliferation, immune dysregulation, and skin barrier impairment. Given that the main pathogenic drivers are fungal proliferation and inflammation, first-line therapies typically involve topical antifungal (e.g., ketoconazole, ciclopirox, miconazole), anti-inflammatory (corticosteroids) and calcineurin inhibitory agents (e.g., pimecrolimus, tacrolimus). Additional options include lithium gluconate/succinate, coal tar, benzoyl peroxide, and phototherapy, among others. Systemic treatments are also generally reserved for severe or resistant cases, with oral terbinafine and fluconazole being the most common (Borda et al., 2019; Dall'oglio et al., 2022; Gupta et al., 2014).

Chitosan shows strong potential in managing seborrheic dermatitis, its antimicrobial action helps control *Malassezia* (H.-F. Huang & Peng, 2015) and *Staphylococcus aureus* (Asadpoor et al., 2021), restoring microbial balance. Chitosan's anti-inflammatory properties modulate key pathways such as cGAS-STING and NLRP3, reducing cytokine release and preventing flare-ups (Zosangpuii & Sudheer, 2024). It also regulates sebaceous gland activity, limiting excessive sebum, and forms a protective film that enhances barrier function, prevents moisture loss, and provides antioxidant protection (Khubiev et al., 2023). Together, these multifunctional effects highlight chitosan as a promising therapeutic and preventive option for seborrheic dermatitis, underlining the need for further clinical investigation (Y. Liu et al., 2025).

In the case of acnes vulgaris, the inflammatory state can be triggered by skin microbiota dysbiosis, which is, characterized by a hyperproliferation of *Cutibacterium acnes* under specific conditions (e.g., excessive sebum production), highlighting the relevance of skin environment - skin microbiota interactions on the onset of this type of disease (Vasam et al., 2023). Topical therapy commonly includes retinoids (e.g., tretinoin) and antibiotics (e.g., erythromycin), while combination treatments often incorporate benzoyl peroxide, salicylic acid, niacinamide, azelaic acid and dapsone. Systemic therapy is generally reserved for cases unresponsive to topical approaches or for patients presenting nodular skin or scarring. These treatments likewise rely on retinoids (e.g., isotretinoin) and a broad range of antibiotics (e.g., erythromycin, clindamycin). In female patients, hormonal approaches such oral contraceptives are also used (Reynolds et al., 2024; Vasam et al., 2023).

A randomized double-blind clinical trial revealed that spironolactone loaded chitosan-coated nano lipid carriers effectively treated mild to moderate acne vulgaris, leading to improved healing and reduced lesion count after 8 weeks of therapy compared to the placebo. Also, the nano carriers successfully addressed both non-inflammatory and inflammatory lesions without adverse effects on the skin. Further, the formulation proved to be efficient, presenting encapsulation efficiency of 79.88%, while allowing a sustained drug-release profile when exploring *in vitro* skin absorption models (Saeedi et al., 2024). In another research work, melittin-loaded chitosan nanofibers were shown to inhibit *C. acnes* growth *in vitro*, as well as to significantly attenuate inflammation (ear thickness) in an *in vivo C. acnes* subcutaneous injection-induced acne vulgaris mouse model. Further *in vitro* experiments also revealed that the nanofibers were non-toxic on human dermal fibroblasts. In all cases, the incorporation of chitosan-based nanofibers allowed the improvement of the results obtained with plain melittin. Additionally, the formulation was able to load up to 86.74% of the targeted drug and allowed the release of 89.65% of its content over 72 h (Rahnama et al., 2022).

4.2.2. Tissue Regeneration and Wound Healing

As previously described, wounds resulting from trauma, burns, surgical interventions or chronic diseases affect millions of individuals worldwide, causing pain and discomfort and in severe cases, even leading to disability (Ding et al., 2024). For instance, chronic wounds represent a major financial burden for healthcare systems worldwide. In the United States of America (USA), it is estimated that around 2% of the population suffers from some form of chronic wound. Thus, wound treatment strategies must be both effective and economically sustainable to help lower mortality rates and ease the financial burden. While acute wounds typically progress through the natural healing stages, chronic or hard-to-heal wounds - such as those caused by severe burns, large tissue loss, nonhealing ulcers or diabetes - are much more complex. Surgical approaches to wound management include procedures like skin grafting and flap surgeries, whereas non-surgical options rely on topical agents, specialized dressings, and skin substitutes (Kolimi et al., 2022; Pereira & Bártolo, 2014). Chitosan-based biomaterials promote wound healing through a variety of mechanisms. Silver nanoparticle cryogels improved hemostasis and antibacterial effects, whilst nanofibers stimulated regeneration and fibroblast proliferation. Crisaborole-loaded nanoemulgels had both anti-inflammatory and healing properties, while vascular endothelial growth factor (VEGF)-loaded sponges enhanced angiogenesis and wound repair in diabetic wounds.

For instance, a shape memory chitosan-based cryogel incorporating silver nanoparticles was developed for wound repair purposes. Comparing the formulation to commercial gelatine sponges and gauze, hemostasis tests showed that it could absorb a significant amount of blood and encourage blood cell adhesion, which in turn promoted blood coagulation. Additionally, the cryogel exerted strong antibacterial activity over *S. aureus* and *Escherichia coli* and significantly promoted wound healing in an *in vivo* model of full-thickness wounds infected with *S. aureus* in Sprague-Dawley rats (Cao et al., 2023). Furthermore, chitosan-based electrospun nanofibers derived from *Periplaneta americana* residues exhibited excellent mechanical and antibacterial properties, and biocompatibility while facilitating wound healing in an infected rat whole-layer wound model by encouraging wound closure, epithelialization, collagen deposition, and inflammation reduction (X. Li et al., 2023). Similarly, in another study, chitosan-based nanofibers were developed, in this case, incorporating antibacterial silver and zinc oxide nanoparticles. The results demonstrated that the nanocomposites exhibited a high antioxidant effect and antibacterial activity against *S. aureus*, *E. coli*, and *Pseudomonas aeruginosa*. Further, *in vitro* experiments showed that the formulations were nontoxic to fibroblasts, while showing good blood compatibility, as well as significantly aiding fibroblast migration and proliferation during a wound healing assay (scratching method), thus enhancing skin tissue regeneration (Bagheri et al., 2022). A different study evaluated the formulation, physicochemical characterization, and wound healing properties of a crisaborole-loaded chitosan-based nanoemulgel. The optimized formulation showed promising wound healing activity in an *in vivo* excision wound Wistar rat model, and potent anti-inflammatory action by inhibiting carrageenan-induced paw edema formation, thereby suggesting that the crisaborole-loaded chitosan-based nanoemulgel can be effectively used to treat skin wounds (Ansari et al., 2022). Finally, the use of chitosan-hyaluronic acid composite sponges with VEGF-loaded fibrin nanoparticles to enhance angiogenesis in diabetic wounds was studied. The study's results revealed that the sponges were efficient for the sustained release of VEGF, with over 60% of the content being released in three days, while improving endothelial cell proliferation, enabling the formation of capillary-like structures in HUVECs. These findings suggest potential improvements in the wound healing process (Mohandas et al., 2015).

4.2.3. Skin Cancer

Skin cancer represents a heterogeneous group of diseases and a worldwide epidemic. Broadly, it can be classified into melanoma and non-melanoma skin cancers, the latter accounting for more than 90% of all cases (Roky et al., 2025). The optimal treatment strategy for skin cancer depends on the tumour's type, size, region and developmental stage. In early stages, standard approaches include excision surgery, Mohs surgery, and radiation therapy for large-sized skin cancer, while smaller sites are often managed with curettage and electrodesiccation, cryotherapy, laser therapy and photodynamic therapy. In advanced or metastatic disease, where surgery or radiotherapy are not feasible, immunotherapy, targeted therapy, and chemotherapy are the main options. These treatments are also applied after surgery or radiotherapy to reduce the risk of recurrence (Hasan et al., 2023; Zeng et al., 2023). Chitosan-based systems show promise in treating skin cancer. In melanoma cell and animal models, drug-loaded nanoparticles, chitosan-coated liposomes, and functionalized nanomicelles increased cytotoxicity, skin permeation, tumor suppression, and survival rates. For non-melanoma malignancies, chitosan-encapsulated silver nanoparticles and lipid-coated nanocarriers improved anticancer activity and photodynamic treatment efficacy while enhancing drug retention in the skin.

Concerning melanoma treatment, one study reported the development of 10-hydroxycamptothecin-loaded nanoparticles to improve chemotherapy. Regarding their anti-melanoma activity, *in vitro* experiments revealed that the loaded-nanoparticles induced apoptosis in B16F10 cells and showed greater cytotoxicity over the same cells in comparison to free 10-hydroxycamptothecin. Further, in an *in vivo* B16F10 cell-based melanoma mouse model, the loaded-nanoparticles were also able to significantly inhibit the progression of tumors (Guo et al., 2020). In another study, chitosan-coated liposomes were developed as a formulation strategy to stabilize and enhance skin permeation of indocyanine green for topical melanoma photodynamic therapy. Results showed that chitosan-coated liposomes protected indocyanine green from degradation while uncoated liposomes did not. Moreover, they significantly increased cellular uptake and photocytotoxicity of indocyanine green in B16F10 melanoma cells in a chitosan-dependent manner. The skin permeation of indocyanine green was also drastically improved by chitosan-coated liposomes (E. H. Lee et al., 2019). Additionally, the development of functionalized thiolated chitosan-based nanomicelles for ergotamine delivery in melanoma treatment has been reported. Characterization studies revealed that the functionalized nanomicelles demonstrated improved physicochemical properties, skin adhesion, and mucopermeation when compared to unmodified polymers. Also, they showed higher release rates of ergotamine at pH 5.5, which relates to acidified conditions of the tumor microenvironment. In an *in vivo* 7,12-dimethylbenz[a] anthracene-induced melanoma mouse model, the administration of the functionalized nanomicelles resulted in reduced tumor size and weight, concomitant with improvements in the survival rate (Azhar et al., 2025).

Regarding non-melanoma cancers, a study reported the development of *Achillea millefolium* extract-loaded silver nanoparticles (Ag NPs) encapsulated by chitosan polymers, which exhibited anti-cutaneous squamous cell carcinoma effects by compromising the viability of several cancerous cell lines (PM1, MET1, MET 4, SCC T9, SCCIC1MET, SCC IC19, SCC T8, and SCC T11), presenting IC_{50} values ranging from 62 to 183 μg/mL. Also, the exerted toxicity over these cells was concentration-dependent (Zheng et al., 2024). Furthermore, lipid-coated chitosan nanocarriers for skin delivery of ferrous chlorophyllin as a potential treatment of squamous cell carcinoma by photodynamic therapy have been explored. In this study, these formulations were shown to be toxic and cause cell disintegration after laser exposure against human squamous cell carcinoma monolayer and three-dimensional spheroids. Also, the chitosan-based nanocarriers presented an entrapment efficiency of 68%, and significantly improved drug-retention in the skin when compared to the ferrous chlorophyllin plain solution (Nasr, Rady, Gomaa, Syrovets, et al., 2019).

5. Chitosan in Metabolic Health and Diabetes

5.1. Metabolic Illnesses: Pathophysiological Background

Metabolic diseases represent a growing global health challenge, contributing substantially to the burden of non-communicable diseases. These disorders encompass a range of interconnected metabolic dysregulations involving energy balance, glucose and lipid homeostasis, and inflammatory pathways. Conditions such as obesity, type 2 diabetes mellitus (T2DM), hypertension, hyperlipidemia, and non-alcoholic fatty liver disease fall under this spectrum. These diseases often coexist and share common underlying risk factors (poor diet, sedentary lifestyle, and genetic predisposition) collectively referred to as metabolic syndrome (Chew et al., 2023).

Metabolic syndrome is not a single disease but a cluster of metabolic abnormalities, including abdominal obesity, dyslipidemia, impaired glucose regulation, and elevated blood pressure, significantly increasing the risk of cardiovascular disease and T2DM (Al-Okbi et al., 2023). In 2009, a consensus statement (Alberti et al., 2009) by the International Diabetes Federation and the American Heart Association/National Heart, Lung, and Blood Institute standardized the definition, stating that metabolic syndrome is diagnosed when three or more of the following criteria are present. The criteria include a waist circumference greater than 101.6 cm in men or 88.9 cm in women, fasting blood glucose levels above 100 mg/dL, triglyceride concentrations higher than 150 mg/dL, HDL cholesterol levels below 40 mg/dL in men or 50 mg/dL in women, and blood pressure exceeding 130/85 mmHg.

The presence of metabolic syndrome substantially heightens the risk of cardiovascular disease, certain cancers, disability, and premature mortality. The global prevalence of both metabolic syndrome and related diseases continues to rise, presenting profound clinical, social, and economic challenges (Chew et al., 2023). Current management primarily relies on lifestyle modification and pharmacological intervention. While dietary changes, physical activity, and weight loss remain the cornerstone strategies, they are often insufficient to achieve long-term control. Drug therapies, although effective, may cause adverse effects such as gastrointestinal disturbances, hypoglycemia, or muscle pain, and their long-term use can be costly. These limitations have prompted growing interest in alternative approaches, particularly natural bioactive compounds. Among these, chitosan has emerged as a promising candidate due to its biodegradability, non-toxicity, and hypoallergenic properties (Tao et al., 2021).

5.2. Chitosan and Pancreatic β-Cells Protection

Hyperglycemia plays a key role in mediating most of the diabetic complications and impaired cell function in T2DM. Combined with irreversible β-cell impairment, this condition increases glucose toxicity, which further worsens the disease by disrupting β-cell function (Priyanka et al., 2022). As the levels of reducing sugars increase in the blood, a set of reactions are triggered that ultimately result in the formation of reactive oxygen species (ROS), promoting damage induced by oxidative stress (Karadeniz et al., 2010). Pancreatic β-cells are particularly prone to these reactive species due to their innate decreased expression of anti-oxidative enzymes, which can damage their DNA, lipids, and proteins (Priyanka et al., 2022). Although patients may maintain healthy β-cells for years after T2DM onset, prolonged exposure to increased glucose levels will eventually damage β-cell function in a later stage of the disease (Karadeniz et al., 2010). Research suggests that antioxidants found in many dietary sources can protect β-cells against free radicals which lead to apoptosis and glucotoxicity (Karadeniz et al., 2010). According to Tzeng et al., (2022), COS (chitosan oligosaccharide) protects pancreatic β-cells through multiple mechanisms. First, COS exhibits strong antioxidant activity by scavenging ROS and reducing oxidative stress, thereby preventing cellular damage. Second, COS enhances the activity of endogenous antioxidant enzymes, including superoxide dismutase (SOD) and catalase (CAT), which play crucial roles in mitigating oxidative injury. Third, COS decreases lipid peroxidation and inhibits apoptosis in β-cells, preserving their structural integrity and functional capacity. Additionally, COS helps maintain normal insulin secretion by improving β-cell viability under hyperglycemic conditions. Collectively, these actions contribute to the protection and restoration of β-cell function, which is essential for improving glucose homeostasis in T2DM (Tzeng et al., 2022).

Yuan et al., (2009) further demonstrated that the β-cell protective effects of COS are strongly influenced by their molecular characteristics, with low molecular weight and highly deacetylated COS exhibiting the greatest activity. Their study showed that COS significantly prevented DNA fragmentation and oxidative DNA damage in pancreatic β-cells exposed to streptozotocin (STZ), as confirmed by DNA ladder and SCGE assays. COS supplementation also preserved islet morphology by reducing β-cell loss, nuclear pyknosis, and islet shrinkage in STZ-induced diabetic rats. Additionally, COS enhanced total antioxidant capacity (T-AOC) and reduced malondialdehyde (MDA) levels *in vivo*, correlating with better structural protection of pancreatic tissue. These findings indicate that COS provides robust β-cell protection by maintaining genomic integrity, reducing apoptosis, and preserving islet architecture under oxidative stress conditions (Yuan et al., 2009).

Chronic hyperglycemia in T2DM induces oxidative stress and progressive β-cell dysfunction, establishing a detrimental cycle that promotes disease advancement. Studies conducted *in vitro* and *in vivo* demonstrate the efficacy of COS in mitigating these effects by scavenging ROS, augmenting endogenous antioxidant defenses, inhibiting apoptosis, and maintaining β-cell structure and function. Their activity seems to be affected by molecular characteristics like molecular weight and the degree of deacetylation, which shows how important it is to optimize the structure for therapeutic effectiveness. These findings indicate that COS are promising candidates for β-cell protection and glucose regulation in T2DM, necessitating further exploration of their mechanisms and translational potential.

5.3. Chitosan and Effects on Glucose Homeostasis

Multiple studies have provided evidence in favor of chitosan and its derivatives' ability to decrease glucose in a model of diabetic mice. Both the up-regulation and down-regulation of several important molecules involved in the glucose-regulating pathways in the liver, muscle, and adipocytes are part of chitosan's anti-hyperglycemic activity. Numerous types of chitosan with varying sizes and molecular weights have been tested to investigate their potential to improve diabetes mellitus (Sarkar et al., 2020). Building on this, Tzeng et al., (2022) report that COS exert significant effects on glucose metabolism by improving insulin sensitivity and enhancing glucose utilization. COS reduces hepatic glucose production, thereby lowering endogenous glucose output, and simultaneously increases glucose uptake in peripheral tissues, particularly skeletal muscle, through upregulation of glucose transporter proteins. This dual action helps to maintain normoglycemia under hyperglycemic conditions. Additionally, COS has been shown to modulate key metabolic enzymes involved in glycolysis and gluconeogenesis, further supporting improved glucose homeostasis. These mechanisms collectively suggest that COS can play a vital role in regulating blood glucose levels and alleviating hyperglycemia in T2DM. Furthermore, S. H. Liu et al., (2010) demonstrated that both high and low molecular weight chitosan enhanced glucose uptake in skeletal muscle, an effect likely mediated by increased GLUT4 translocation and enhanced Akt phosphorylation in STZ-induced diabetic rats. In addition, chitosan reduced hepatic phosphorylation of the p38 protein and suppressed the expression of phosphoenolpyruvate carboxykinase (PEPCK). As a key regulatory enzyme in gluconeogenesis, PEPCK catalyzes an irreversible step critical for maintaining hepatic glucose homeostasis. Hence, its downregulation contributes to improved glycemic control (Sarkar et al., 2020).

In addition to controlling PEPCK expression, adenosine monophosphate (AMP)-activated protein kinase (AMPK) is a key modulator of cellular energy balance (Inoue & Yamauchi, 2006). S. H. Liu et al., (2010) studied into whether chitosan affected the AMPK pathway to understand this process. They found that after treatment, AMPK phosphorylation significantly increased in diabetic rats. This discovery highlights a potential therapeutic approach for the treatment of diabetes by indicating that chitosan may enhance glycemic control by concurrently activating AMPK and inhibiting PEPCK (Sarkar et al., 2020). Finally, a recent randomized, double-blind, controlled crossover trial evaluated the hypoglycemic efficacy of COS supplementation in individuals with impaired glucose tolerance, impaired fasting glucose, and healthy controls. COS supplementation (500 mg low-molecular weight COS per dose with meals) significantly reduced postprandial blood glucose levels compared with placebo (egg white powder). However, the study was limited by its small sample size and the absence of insulin and C-peptide measurements. Furthermore, the glycemic benefits of COS require validation in larger trials, particularly among diabetic patients (Sarkar et al., 2020).

When combined, these results show that chitosan and its derivatives are prospective multi-target medicines for the treatment of T2DM. Chitosan has a wide range of therapeutic possibilities for re-establishing glucose homeostasis by boosting skeletal muscle glucose absorption, optimizing insulin sensitivity, decreasing hepatic glucose production, and controlling important metabolic enzymes via AMPK activation. These benefits are repeatable across animal models and chitosan forms, highlighting its potential as a supplemental diabetes therapy approach that merits more mechanistic research and clinical assessment.

5.4. Chitosan-Based Delivery Systems for Antidiabetic and Lipid Absorption Agents

Chitosan-based delivery systems have tremendous promise in antidiabetic treatment. Their capacity to encapsulate a wide range of compounds, including hormones, proteins, and tiny molecules, enables their regulated and prolonged release within the body. By guaranteeing that medications are administered in the appropriate dosages at the appropriate times, controlled release enhances drug effectiveness and produces better treatment results. Chitosan's adaptability also makes it a potential substance for administering medications that are poorly soluble or that need to be shielded from the digestive tract to continue working (Ghazwani et al., 2025).

Many antidiabetic drugs, such as glibenclamide and pioglitazone, have poor water solubility, while protein-based therapies like insulin are easily degraded by stomach acid and digestive enzymes (Confederat et al., 2020). Chitosan-based delivery systems offer a powerful solution to these problems.

Firstly, by encapsulating therapeutic agents within a polymeric matrix, chitosan shields them from the harsh, low-pH environment of the stomach and protects them from enzymatic degradation throughout the gastrointestinal (GI) tract (Lupascu et al., 2024). Furthermore, the polymer's positive charge enables it to adhere to the negatively charged mucosal lining of the intestine. This prolongs the residence time of the drug at the site of absorption, thereby increasing its bioavailability. Chitosan can also transiently and reversibly open the tight junctions between intestinal epithelial cells, allowing drugs to pass through the paracellular pathway and enter the bloodstream more effectively (Alfatama et al., 2024). Finally, chitosan matrices can be engineered to provide controlled or sustained release of the encapsulated drug, which helps maintain stable therapeutic concentrations and improves patient compliance (Confederat et al., 2020). With regards to notable applications, researchers have successfully formulated chitosan into nanoparticles (NPs) and microparticles (MPs) for delivering a range of antidiabetic agents. Confederat et al., (2020) developed a system of chitosan microparticles to co-deliver the antidiabetic drug glibenclamide and the antioxidant lipoic acid. The formulation demonstrated high entrapment efficiency and a favorable, controlled release profile in simulated GI fluids, highlighting its potential as a multi-target therapy for diabetes. Furthermore, chitosan is also one of the most widely researched polymers for developing an oral insulin formulation. Chitosan-based NPs have been shown to protect insulin from degradation while its mucoadhesive and permeation-enhancing properties facilitate its transport across the intestinal barrier (Alfatama et al., 2024). Furthermore, chitosan's primary mechanism for weight management stems from its unique chemical properties within the GI tract. In the acidic environment of the stomach, the amino groups in chitosan's structure become protonated, conferring a positive charge to the polymer (Fatahi et al., 2022). This cationic nature allows it to electrostatically bind with negatively charged molecules like fatty acids and bile acids, forming large, indigestible complexes. These complexes are subsequently excreted, effectively reducing dietary fat absorption and promoting weight loss (Moraru et al., 2018).

The efficacy of this mechanism can vary based on chitosan's molecular weight High-molecular-weight chitosan (HC) primarily relies on this physical binding to increase fecal excretion of fats and bile acids. In contrast, chitosan oligosaccharides (CO), which are water-soluble and more easily absorbed, exert their effects differently by downregulating the expression of intestinal genes responsible for lipid absorption, such as fabp2 and fatp4 (S.-H. Liu et al., 2021). Finally, clinical evidence supports chitosan's role as an adjuvant treatment for obesity. Meta-analyses of randomized controlled trials (RCTs) have shown that chitosan supplementation results in a modest but statistically significant reduction in body weight and body mass index (BMI) compared to a placebo (Moraru et al., 2018). For instance, one meta-analysis found an average weight loss of -1.01 kg among overweight and obese patients. Studies also demonstrate improvements in lipid profiles, including significant reductions in total cholesterol, LDL cholesterol, and triglycerides (Kholdebarin et al., 2024). An RCT conducted on adolescents with obesity confirmed these findings, reporting significant improvements not only in weight and BMI but also in lipid profiles and appetite-regulating hormones like leptin and adiponectin after 12 weeks of supplementation. While not a standalone solution, chitosan is an effective component of a comprehensive weight management program when combined with caloric restriction and physical activity (Fatahi et al., 2022).

Chitosan-based systems represent a significant advancement in the management of diabetes and obesity. It is a particularly important biopolymer because of its dual roles as a sophisticated carrier for antidiabetic medications and a natural inhibitor of fat absorption. Chitosan offers a multimodal approach to improving metabolic health and provides a viable foundation for the development of more effective and patient-friendly medicines by lowering calorie intake from lipids while also boosting the oral bioavailability and efficacy of essential pharmaceuticals.

Conclusion

Chitosan has demonstrated broad potential across biomedical fields, with a consistent set of structural and functional features supporting its adaptability. Its cationic nature, modifiable chemical groups, and biodegradability provide a foundation for applications in dialysis, dermatology, and diabetes, each of which leverages different aspects of its chemistry. In dialysis, chitosan offers improved hemocompatibility and adsorption of protein-bound toxins when integrated into synthetic membranes. In dermatology, it contributes antimicrobial, anti-inflammatory, and regenerative effects, enhancing topical treatments for inflammatory diseases, wounds, and cancer. In diabetes and metabolic health, COS exhibit protective actions on β-cells, improve glucose homeostasis, and act as carriers for antidiabetic drugs and lipid-lowering therapies. Together, these findings reinforce the idea that chitosan is not a single-purpose material, but a platform technology with cross-cutting biomedical value.

Recent trends reveal a movement toward fine-tuned molecular and nanoscale modifications that expand functionality beyond the native polymer. Strategies such as quaternization and sulfonation enhance solubility and selectivity, while nanocomposite approaches incorporating metals, oxides, or graphene derivatives add antimicrobial and antioxidant capacity. Electrospinning has enabled the creation of nanofiber membranes that mimic extracellular matrix architecture, offering high porosity and surface area. Advances in green extraction methods and fungal-derived chitosan are redefining the sustainability profile of the material, with increasing attention to circular bioeconomy principles and the reduction of marine-based allergen risks. These innovations mark a shift from chitosan as a bulk polymer to a highly engineered, application-specific biomaterial.

Nevertheless, significant research gaps remain. The variability in degree of deacetylation, molecular weight, and purity across sources and production methods continues to limit reproducibility and comparability between studies. Long-term *in vivo* performance, particularly under repeated or chronic exposure as in dialysis, remains insufficiently characterized. The scalability of green processing technologies is still underdeveloped, creating a tension between sustainability goals and industrial feasibility. There is also a lack of well-designed clinical trials that can validate preclinical findings and provide robust evidence for regulatory approval. Suggested pathways forward include harmonizing characterization methods, investing in pilot-scale green production systems, and prioritizing translational studies that move beyond proof-of-concept toward clinical evaluation.

Chitosan exemplifies the potential of natural polymers to combine biofunctionality, sustainability, and versatility in biomedical applications. Its comparative advantages over conventional synthetic materials are increasingly recognized, but its translation into routine clinical practice depends on bridging gaps in standardization, scalability, and long-term validation. By advancing molecular engineering, green processing, and translational research in parallel, the field can unlock chitosan's full potential and establish it as a cornerstone of next-generation biomedical technologies that are both effective and environmentally responsible.

Acknowledgments

We would like to thank the scientific collaboration of CBQF under the FCT projectUID/Multi/50016/2025. The authors Maria Martingo and Maria Leonor Castro are also grateful to FCT for the grants 2024.03874.BDANA and 2024.03878.BDANA, respectively.

References

Abasian, L., Shafiei Alavijeh, R., Satari, B., & Karimi, K. (2020). Sustainable and Effective Chitosan Production by Dimorphic Fungus Mucor rouxii via Replacing Yeast Extract with Fungal Extract. *Applied Biochemistry and Biotechnology*, *191*(2), 666–678. https://doi.org/10.1007/s12010-019-03220-w.

Ahmed, S. F., Mofijur, M., Rafa, N., Chowdhury, A. T., Chowdhury, S., Nahrin, M., Islam, A. B. M. S., & Ong, H. C. (2022). Green approaches in synthesising nanomaterials for environmental nanobioremediation: Technological advancements, applications, benefits and challenges. *Environmental Research*, *204*. https://doi.org/10.1016/j.envres.2021.111967.

Alberti, K. G. M. M., Eckel, R. H., Grundy, S. M., Zimmet, P. Z., Cleeman, J. I., Donato, K. A., Fruchart, J. C., James, W. P. T., Loria, C. M., & Smith, S. C. (2009). Harmonizing the metabolic syndrome: A joint interim statement of the international diabetes federation task force on epidemiology and prevention; National heart, lung, and blood institute; American heart association; World heart federation; International atherosclerosis society; And international association for the study of obesity. In *Circulation* (Vol. 120, Issue 16). https://doi.org/10.1161/CIRCULATIONAHA.109.192644.

Alemu, D., Getachew, E., & Mondal, A. K. (2023). Study on the Physicochemical Properties of Chitosan and their Applications in the Biomedical Sector. *International Journal of Polymer Science*, *2023*(1), 5025341. https://doi.org/https://doi.org/10.1155/2023/5025341.

Alfatama, M., Choukaife, H., Alkhatib, H., Al Rahal, O., & Zin, N. Z. M. (2024). A comprehensive review of oral chitosan drug delivery systems: Applications for oral insulin delivery. In *Nanotechnology Reviews* (Vol. 13, Issue 1). https://doi.org/10.1515/ntrev-2023-0205.

Almeida, P., Pires, R. F., Brilhante, D., Macêdo, A., Bonifácio, V. D. B., & Faria, M. (2025). Enhanced protein-bound uremic toxin clearance by a membrane-drug-dendrimer synergic system. *Separation and Purification Technology*, *359*. https://doi.org/10.1016/j.seppur.2024.130519.

Al-Okbi, S. Y., Ali, O., Aly, A. S., Refaat, D., Esmail, R. S. H., & Elbakry, H. F. H. (2023). Management of metabolic syndrome by nutraceuticals prepared from chitosan and ferulic acid with or without beta-sitosterol and their nanoforms. *Scientific Reports*, *13*(1). https://doi.org/10.1038/s41598-023-38837-9.

Anraku, M., Tomida, H., Michihara, A., Tsuchiya, D., Iohara, D., Maezaki, Y., Uekama, K., Maruyama, T., Otagiri, M., & Hirayama, F. (2012). Antioxidant and renoprotective activity of chitosan in nephrectomized rats. *Carbohydrate Polymers*, *89*(1), 302–304. https://doi.org/10.1016/j.carbpol.2012.03.014.

Ansari, M. N., Soliman, G. A., Rehman, N. U., & Anwer, M. K. (2022). Crisaborole Loaded Nanoemulsion Based Chitosan Gel: Formulation, Physicochemical Characterization and Wound Healing Studies. *Gels*, *8*(5). https://doi.org/10.3390/gels8050318.

Aranaz, I., Alcántara, A. R., Civera, M. C., Arias, C., Elorza, B., Caballero, A. H., & Acosta, N. (2021). Chitosan: An overview of its properties and applications. In *Polymers* (Vol. 13, Issue 19). MDPI. https://doi.org/10.3390/polym13193256.

Arias-Guillén, M., Martínez Cadenas, R., Gómez, M., Martín Vaquero, N., Pereda, G., Audije-Gil, J., Portillo, J., Quintela, M., Castaño, I., Luque, A., Maduell, F., Ortiz, A., Duane, B., & Arenas, M. D. (2024). Environmental challenges in hemodialysis: Exploring the road to sustainability. In *Nefrologia*. Elsevier Espana S.L.U. https://doi.org/10.1016/j.nefro.2024.06.001.

Asadpoor, M., Ithakisiou, G. N., van Putten, J. P. M., Pieters, R. J., Folkerts, G., & Braber, S. (2021). Antimicrobial Activities of Alginate and Chitosan Oligosaccharides Against Staphylococcus aureus and Group B Streptococcus. *Frontiers in Microbiology*, *12*. https://doi.org/10.3389/fmicb.2021.700605.

Ávila, E., Sepúlveda, R. A., Retamal, J., & Hachim, D. (2025). Biocompatibility in hemodialysis: artificial membrane and human blood interactions. In *BMC Nephrology* (Vol. 26, Issue 1). BioMed Central Ltd. https://doi.org/10.1186/s12882-025-04401-y.

Azhar, F., Naureen, H., Shahnaz, G., Hamdani, S. D. A., Shareef, U., Khattak, S., Babar, M. M., Rahdar, A., Fathi-karkan, S., & Pandey, S. (2025). Chitosan Functionalized Skin-Adhesive Formulation for Ergotamine Delivery in Melanoma Treatment. *BioNanoScience*, *15*(1). https://doi.org/10.1007/s12668-024-01638-z.

Backston, K. J., Thompson, C. J., Raina, M., Sathik, N. R., Noh, L. S., Lokkur, P. P., & Bansal, S. B. (2025). Green Dialysis Review: Reducing Water, Energy, and Waste. *Hemodialysis International*, *n/a*(n/a). https://doi.org/https://doi.org/10.1111/hdi.13251.

Bagheri, M., Validi, M., Gholipour, A., Makvandi, P., & Sharifi, E. (2022). Chitosan nanofiber biocomposites for potential wound healing applications: Antioxidant activity with synergic antibacterial effect. *Bioengineering and Translational Medicine*, *7*(1). https://doi.org/10.1002/btm2.10254.

Baroudi, A., García-Payo, C., & Khayet, M. (2023). Chitosan-Based Composite Membranes with Different Biocompatible Metal Oxide Nanoparticles: Physicochemical Properties and Drug-Release Study. *Polymers*, *15*(13). https://doi.org/10.3390/polym15132804.

Bhatt, P., Joshi, S., Urper Bayram, G. M., Khati, P., & Simsek, H. (2023). Developments and application of chitosan-based adsorbents for wastewater treatments. *Environmental Research*, *226*. https://doi.org/10.1016/j.envres.2023.115530.

Bhattarai, N., Gunn, J., & Zhang, M. (2010). Chitosan-based hydrogels for controlled, localized drug delivery. In *Advanced Drug Delivery Reviews* (Vol. 62, Issue 1, pp. 83–99). https://doi.org/10.1016/j.addr.2009.07.019.

Bieber, T. (2023). Disease modification in inflammatory skin disorders: opportunities and challenges. *Nature Reviews Drug Discovery*, *22*(8), 662–680. https://doi.org/10.1038/s41573-023-00735-0.

Blagg, C. R. (2007). The Early History of Dialysis for Chronic Renal Failure in the United States: A View From Seattle. *American Journal of Kidney Diseases*, *49*(3), 482–496. https://doi.org/10.1053/j.ajkd.2007.01.017.

Borda, L. J., Perper, M., & Keri, J. E. (2019). Treatment of seborrheic dermatitis: a comprehensive review. *Journal of Dermatological Treatment*, *30*(2), 158–169. https://doi.org/10.1080/09546634.2018.1473554.

Bowry, S. K., Kircelli, F., Himmele, R., & Nigwekar, S. U. (2021). Blood-incompatibility in haemodialysis: Alleviating inflammation and effects of coagulation. *Clinical Kidney Journal*, *14*, I59–I71. https://doi.org/10.1093/ckj/sfab185.

Calabrese, G., Licata, G., Gambardella, A., De Rosa, A., Alfano, R., & Argenziano, G. (2022). Topical and Conventional Systemic Treatments in Atopic Dermatitis: Have They Gone Out of Fashion? *Dermatology Practical and Conceptual*, *12*(1). https://doi.org/10.5826/dpc.1201a155.

Cao, S., Bi, Z., Li, Q., Zhang, S., Singh, M., & Chen, J. (2023). Shape memory and antibacterial chitosan-based cryogel with hemostasis and skin wound repair. *Carbohydrate Polymers*, *305*, 120545. https://doi.org/https://doi.org/10.1016/j.carbpol.2023.120545.

Castro, M. L., Ferreira, J. P., Pintado, M., Ramos, O. L., Borges, S., & Baptista-Silva, S. (2023). Grape By-Products in Sustainable Cosmetics: Nanoencapsulation and Market Trends. In *Applied Sciences (Switzerland)* (Vol. 13, Issue 16). Multidisciplinary Digital Publishing Institute (MDPI). https://doi.org/10.3390/app13169168.

Castro, R., Guerrero-Legarreta, I., & Bórquez, R. (2018). Chitin extraction from Allopetrolisthes punctatus crab using lactic fermentation. *Biotechnology Reports*, *20*. https://doi.org/10.1016/j.btre.2018.e00287.

Chatterjee, S., Adhya, M., Guha, A. K., & Chatterjee, B. P. (2005). Chitosan from Mucor rouxii: production and physico-chemical characterization. *Process Biochemistry*, *40*(1), 395–400. https://doi.org/https://doi.org/10.1016/j.procbio.2004.01.025.

Chew, N. W. S., Ng, C. H., Tan, D. J. H., Kong, G., Lin, C., Chin, Y. H., Lim, W. H., Huang, D. Q., Quek, J., Fu, C. E., Xiao, J., Syn, N., Foo, R., Khoo, C. M., Wang, J. W., Dimitriadis, G. K., Young, D. Y., Siddiqui, M. S., Lam, C. S. P., … Muthiah, M. D. (2023). The global burden of metabolic disease: Data from 2000 to 2019. *Cell Metabolism*, *35*(3). https://doi.org/10.1016/j.cmet.2023.02.003.

Chou, T. C., Fu, E., Wu, C. J., & Yeh, J. H. (2003). Chitosan enhances platelet adhesion and aggregation. *Biochemical and Biophysical Research Communications*, *302*(3), 480–483. https://doi.org/10.1016/S0006-291X(03)00173-6.

Chuah, L. H., Loo, H. L., Goh, C. F., Fu, J. Y., & Ng, S. F. (2023). Chitosan-based drug delivery systems for skin atopic dermatitis: recent advancements and patent trends. *Drug Delivery and Translational Research*, *13*(5), 1436–1455. https://doi.org/10.1007/s13346-023-01307-w.

Confederat, L. G., Motrescu, I., Condurache, M. I., Constantin, S., Bujor, A., Tuchilus, C. G., & Profire, L. (2020). Chitosan-based delivery systems loaded with glibenclamide and lipoic acid: Formulation, characterization, and kinetic release studies. *Applied Sciences (Switzerland)*, *10*(21). https://doi.org/10.3390/app10217532.

Coulliette, A. D., & Arduino, M. J. (2013). Hemodialysis and water quality. *Seminars in Dialysis*, *26*(4), 427–438. https://doi.org/10.1111/sdi.12113.

Cozzolino, M., Magagnoli, L., & Ciceri, P. (2025). From Physicochemical Classification to Multidimensional Insights: A Comprehensive Review of Uremic Toxin Research. In *Toxins* (Vol. 17, Issue 6). Multidisciplinary Digital Publishing Institute (MDPI). https://doi.org/10.3390/toxins17060295.

Dai, T., Tanaka, M., Huang, Y. Y., & Hamblin, M. R. (2011). Chitosan preparations for wounds and burns: Antimicrobial and wound-healing effects. In *Expert Review of Anti-Infective Therapy* (Vol. 9, Issue 7, pp. 857–879). https://doi.org/10.1586/eri.11.59.

Dall'oglio, F., Nasca, M. R., Gerbino, C., & Micali, G. (2022). An Overview of the Diagnosis and Management of Seborrheic Dermatitis. In *Clinical, Cosmetic and Investigational Dermatology* (Vol. 15, pp. 1537–1548). Dove Medical Press Ltd. https://doi.org/10.2147/CCID.S284671.

Dambuza, A., Mokolokolo, P. P., Makhatha, M. E., & Sibeko, M. A. (2025). Chitosan-Based Materials as Effective Materials to Remove Pollutants. *Polymers*, *17*(18), 2447. https://doi.org/10.3390/polym17182447.

Davis, D., Umesh, M., Santhosh, A. S., Suresh, S., Shanmugam, S., & Kikas, T. (2024). Extraction of Fungal Chitosan by Leveraging Pineapple Peel Substrate for Sustainable Biopolymer Production. *Polymers*, *16*(17). https://doi.org/10.3390/polym16172455.

Desai, N., Rana, D., Salave, S., Gupta, R., Patel, P., Karunakaran, B., Sharma, A., Giri, J., Benival, D., & Kommineni, N. (2023). Chitosan: A Potential Biopolymer in Drug Delivery and Biomedical Applications. In *Pharmaceutics* (Vol. 15, Issue 4). Multidisciplinary Digital Publishing Institute (MDPI). https://doi.org/10.3390/pharmaceutics15041313.

Dhondup, T., & Qian, Q. (2017). Acid-Base and Electrolyte Disorders in Patients with and without Chronic Kidney Disease: An Update. *Kidney Diseases*, *3*(4), 136–148. https://doi.org/10.1159/000479968.

Ding, C., Liu, X., Zhang, S., Sun, S., Yang, J., Chai, G., Wang, N., Ma, S., Ding, Q., & Liu, W. (2024). Multifunctional hydrogel bioscaffolds based on polysaccharide to promote wound healing: A review. In *International Journal of Biological Macromolecules* (Vol. 259). Elsevier B. V. https://doi.org/10.1016/j.ijbiomac.2024.129356.

Du, J. R., Hsu, L. H., Xiao, E. S., Guo, X., Zhang, Y., & Feng, X. (2020). Using genipin as a "green" crosslinker to fabricate chitosan membranes for pervaporative dehydration of isopropanol. *Separation and Purification Technology*, *244*. https://doi.org/10.1016/j.seppur.2020.116843.

Eduok, U., Abdelrasoul, A., Shoker, A., & Doan, H. (2021). Recent developments, current challenges and future perspectives on cellulosic hemodialysis membranes for highly efficient clearance of uremic toxins. In *Materials Today Communications* (Vol. 27). Elsevier Ltd. https://doi.org/10.1016/j.mtcomm.2021.102183.

Elahi, N., Astaneh, M. E., Ai, J., & Rizwan, M. (2024). Atopic dermatitis treatment: A comprehensive review of conventional and novel bioengineered approaches.

International Journal of Biological Macromolecules, *282*, 137083. https://doi.org/https://doi.org/10.1016/j.ijbiomac.2024.137083.

Engelen, M. M., Verhamme, P., & Vanassche, T. (2023). Clotting of the Extracorporeal Circuit in Hemodialysis: Beyond Contact-Activated Coagulation. *Seminars in Nephrology*, *43*(6), 151473. https://doi.org/https://doi.org/10.1016/j.semnephrol.2023.151473.

Fatahi, S., Sayyari, A. A., Salehi, M., Safa, M., Sohouli, M., Shidfar, F., & Santos, H. O. (2022). The effects of chitosan supplementation on anthropometric indicators of obesity, lipid and glycemic profiles, and appetite-regulated hormones in adolescents with overweight or obesity: a randomized, double-blind clinical trial. *BMC Pediatrics*, *22*(1). https://doi.org/10.1186/s12887-022-03590-x.

Feng, W., & Wang, Z. (n.d.). *Biomedical applications of chitosan-graphene oxide nanocomposites*. https://doi.org/10.1016/j.isci.

Feng, X., Zhu, J., Jin, J., Wang, Y., Zhang, Y., & Van der Bruggen, B. (2024). Polymers of intrinsic microporosity for membrane-based precise separations. In *Progress in Materials Science* (Vol. 144). Elsevier Ltd. https://doi.org/10.1016/j.pmatsci.2024.101285.

Fereig, S. A., El-Zaafarany, G. M., Arafa, M. G., & Abdel-Mottaleb, M. M. A. (2021). Tacrolimus-loaded chitosan nanoparticles for enhanced skin deposition and management of plaque psoriasis. *Carbohydrate Polymers*, *268*, 118238. https://doi.org/https://doi.org/10.1016/j.carbpol.2021.118238.

Ghazwani, M., Hani, U., Kyada, A., Ballal, S., Issa, B. S., Abosaoda, M. K., Singh, A., Sabarivani, A., & Ray, S. (2025). Advancements in insulin delivery: the potential of natural polymers for improved diabetes management. *Frontiers in Bioengineering and Biotechnology*, *13*. https://doi.org/10.3389/fbioe.2025.1566743.

Gheorghiță, D., Moldovan, H., Robu, A., Bița, A. I., Grosu, E., Antoniac, A., Corneschi, I., Antoniac, I., Bodog, A. D., & Băcilă, C. I. (2023). Chitosan-Based Biomaterials for Hemostatic Applications: A Review of Recent Advances. In *International Journal of Molecular Sciences* (Vol. 24, Issue 13). Multidisciplinary Digital Publishing Institute (MDPI). https://doi.org/10.3390/ijms241310540.

Gonciarz, W., Balcerczak, E., Brzeziński, M., Jeleń, A., Pietrzyk-Brzezińska, A. J., Narayanan, V. H. B., & Chmiela, M. (2025). Chitosan-based formulations for therapeutic applications. A recent overview. In *Journal of Biomedical Science* (Vol. 32, Issue 1). BioMed Central Ltd. https://doi.org/10.1186/s12929-025-01161-7.

Gortari, M. C., & Hours, R. A. (2013). Biotechnological processes for chitin recovery out of crustacean waste: A mini-review. In *Electronic Journal of Biotechnology* (Vol. 16, Issue 3). https://doi.org/10.2225/vol16-issue3-fulltext-10.

Guo, H., Li, F., Qiu, H., Liu, J., Qin, S., Hou, Y., & Wang, C. (2020). Preparation and Characterization of Chitosan Nanoparticles for Chemotherapy of Melanoma Through Enhancing Tumor Penetration. *Frontiers in Pharmacology*, *11*. https://doi.org/10.3389/fphar.2020.00317.

Gupta, A. K., Richardson, M., & Paquet, M. (2014). Systematic Review of Oral Treatments for Seborrheic Dermatitis. *Journal of the European Academy of Dermatology and Venereology*, *28*(1), 16–26. https://doi.org/https://doi.org/10.1111/jdv.12197.

Hafsa, J., Smach, M. A., Mrid, R. Ben, Sobeh, M., Majdoub, H., & Yasri, A. (2021). Functional properties of chitosan derivatives obtained through Maillard reaction: A novel promising food preservative. In *Food Chemistry* (Vol. 349). Elsevier Ltd. https://doi.org/10.1016/j.foodchem.2021.129072.

Hasan, N., Nadaf, A., Imran, M., Jiba, U., Sheikh, A., Almalki, W. H., Almujri, S. S., Mohammed, Y. H., Kesharwani, P., & Ahmad, F. J. (2023). Skin cancer: understanding the journey of transformation from conventional to advanced treatment approaches. In *Molecular Cancer* (Vol. 22, Issue 1). BioMed Central Ltd. https://doi.org/10.1186/s12943-023-01854-3.

Herdiana, Y., Husni, P., Nurhasanah, S., Shamsuddin, S., & Wathoni, N. (2023). Chitosan-Based Nano Systems for Natural Antioxidants in Breast Cancer Therapy. In *Polymers* (Vol. 15, Issue 13). Multidisciplinary Digital Publishing Institute (MDPI). https://doi.org/10.3390/polym15132953.

Hsu, C. Y., Ajaj, Y., Mahmoud, Z. H., Kamil Ghadir, G., Khalid Alani, Z., Hussein, M. M., Abed Hussein, S., Morad Karim, M., Al-khalidi, A., Abbas, J. K., Hussein Kareem, A., & kianfar, E. (2024). Adsorption of heavy metal ions use chitosan/graphene nanocomposites: A review study. In *Results in Chemistry* (Vol. 7). Elsevier B.V. https://doi.org/10.1016/j.rechem.2024.101332.

Hu, J., Chen, K., Xiang, M., Wei, J., Zeng, Y., Qin, Y., Zhang, L., & Zhang, W. (2024). A novel sponge composite of chitosan-sodium tripolyphosphate-melamine for anionic dye Orange II removal. *International Journal of Biological Macromolecules*, *270*. https://doi.org/10.1016/j.ijbiomac.2024.132056.

Huang, H.-F., & Peng, C.-F. (2015). Antibacterial and antifungal activity of alkylsulfonated chitosan. *Biomarkers and Genomic Medicine*, *7*(2), 83–86. https://doi.org/https://doi.org/10.1016/j.bgm.2014.09.001.

Huang, Y. A., & Lin, W. J. (2025). Nanoparticles composed of polysaccharide chitosan and oligosaccharide alginate for strengthen transdermal delivery of tacrolimus in atopic dermatitis. *Carbohydrate Polymer Technologies and Applications*, *9*, 100655. https://doi.org/https://doi.org/10.1016/j.carpta.2024.100655.

Huq, T., Khan, A., Brown, D., Dhayagude, N., He, Z., & Ni, Y. (2022). Sources, production and commercial applications of fungal chitosan: A review. In *Journal of Bioresources and Bioproducts* (Vol. 7, Issue 2, pp. 85–98). KeAi Communications Co. https://doi.org/10.1016/j.jobab.2022.01.002.

Inoue, E., & Yamauchi, J. (2006). AMP-activated protein kinase regulates PEPCK gene expression by direct phosphorylation of a novel zinc finger transcription factor. *Biochemical and Biophysical Research Communications*, *351*(4). https://doi.org/10.1016/j.bbrc.2006.10.124.

Ji, H., Li, Y., Su, B., Zhao, W., Kizhakkedathu, J. N., & Zhao, C. (2023). Advances in Enhancing Hemocompatibility of Hemodialysis Hollow-Fiber Membranes. In *Advanced Fiber Materials* (Vol. 5, Issue 4, pp. 1198–1240). Springer. https://doi.org/10.1007/s42765-023-00277-5.

Kalantar-Zadeh, K., Ficociello, L. H., Bazzanella, J., Mullon, C., & Anger, M. S. (2021). Slipping through the pores: Hypoalbuminemia and albumin loss during hemodialysis. *International Journal of Nephrology and Renovascular Disease*, *14*, 11–21. https://doi.org/10.2147/IJNRD.S291348.

Karadeniz, F., Artan, M., Kong, C. S., & Kim, S. K. (2010). Chitooligosaccharides protect pancreatic β-cells from hydrogen peroxide-induced deterioration. *Carbohydrate Polymers*, *82*(1). https://doi.org/10.1016/j.carbpol.2010.04.046.

Ke, C. L., Deng, F. S., Chuang, C. Y., & Lin, C. H. (2021). Antimicrobial actions and applications of Chitosan. In *Polymers* (Vol. 13, Issue 6). MDPI AG. https://doi.org/10.3390/polym13060904.

Khanafari, A., Marandi, R., & Sanatei, S. (2008). Recovery of chitin and chitosan from shrimp waste by chemical and microbial methods. *Iranian Journal of Environmental Health Science & Engineering, 5.*

Kholdebarin, M., Pahlavani, N., Nikbaf-Shandiz, M., Mosallaei, H., Rasaei, N., Khalse, Z., Aali, Y., Asbaghi, O., Zamanian, A., & Shiraseb, F. (2024). A Systematic Review and Meta-Analysis to Evaluate the Effects of Chitosan on Obesity Indicators. *Food Science & Nutrition, 12*(12), 10030–10048. https://doi.org/10.1002/fsn3.4596.

Khubiev, O. M., Egorov, A. R., Kirichuk, A. A., Khrustalev, V. N., Tskhovrebov, A. G., & Kritchenkov, A. S. (2023). Chitosan-Based Antibacterial Films for Biomedical and Food Applications. In *International Journal of Molecular Sciences* (Vol. 24, Issue 13). Multidisciplinary Digital Publishing Institute (MDPI). https://doi.org/10.3390/ijms241310738.

Kobayashi, S., & Ohtake, T. (2023). The Characteristics of Dialysis Membranes: Benefits of the AN69 Membrane in Hemodialysis Patients. In *Journal of Clinical Medicine* (Vol. 12, Issue 3). MDPI. https://doi.org/10.3390/jcm12031123.

Kolimi, P., Narala, S., Nyavanandi, D., Youssef, A. A. A., & Dudhipala, N. (2022). Innovative Treatment Strategies to Accelerate Wound Healing: Trajectory and Recent Advancements. In *Cells* (Vol. 11, Issue 15). MDPI. https://doi.org/10.3390/cells11152439.

Kucharska, M., Sikora, M., Brzoza-Malczewska, K., & Owczarek, M. (2019). *Antimicrobial Properties of Chitin and Chitosan* (pp. 169–187). https://doi.org/10.1002/9781119450467.ch7.

Laranjinha, I., Gonçalves, M., & Harber, M. (2023). The (Un)Sustainability of Nephrology. *Portuguese Journal of Nephrology & Hypertension*, *37*(3). https://doi.org/10.32932/pjnh.2023.07.245.

Lee, E.-H., Lim, S.-J., & Lee, M.-K. (2019). Chitosan-coated liposomes to stabilize and enhance transdermal delivery of indocyanine green for photodynamic therapy of melanoma. *Carbohydrate Polymers*, *224*, 115143. https://doi.org/https://doi.org/10.1016/j.carbpol.2019.115143.

Lee, J. S., Oh, E., Oh, H., Kim, S., Ok, S., Sa, J., Lee, J.-H., Shin, Y. C., Bae, Y.-S., Choi, C. Y., Lee, S., Kwon, H.-K., Yang, S., & Choi, W. Il. (2024). Tacrolimus-loaded chitosan-based nanoparticles as an efficient topical therapeutic for the effective treatment of atopic dermatitis symptoms. *International Journal of Biological Macromolecules*, *273*, 133005. https://doi.org/https://doi.org/10.1016/j.ijbiomac.2024.133005.

Li, K., Xing, R., Liu, S., & Li, P. (2016). Advances in preparation, analysis and biological activities of single chitooligosaccharides. In *Carbohydrate Polymers* (Vol. 139, pp. 178–190). Elsevier Ltd. https://doi.org/10.1016/j.carbpol.2015.12.016.

Li, X., Jiang, F., Duan, Y., Li, Q., Qu, Y., Zhao, S., Yue, X., Huang, C., Zhang, C., & Pan, X. (2023). Chitosan electrospun nanofibers derived from Periplaneta americana residue for promoting infected wound healing. *International Journal of Biological Macromolecules*, *229*, 654–667. https://doi.org/https://doi.org/10.1016/j.ijbiomac.2022.12.272.

Liang, J., & Liu, Y. (2023). Animal Models of Kidney Disease: Challenges and Perspectives. In *Kidney360* (Vol. 4, Issue 10, pp. 1479–1493). Lippincott Williams and Wilkins. https://doi.org/10.34067/KID.0000000000000227.

Lin, L., Chen, L., Lu, C., Chen, G., & Hong, F. F. (2024). Chitosan particles embedded bacterial nanocellulose flat membrane for hemodialysis. *International Journal of Biological Macromolecules*, *266*, 130646. https://doi.org/https://doi.org/10.1016/j.ijbiomac.2024.130646.

Liu, S. H., Chang, Y. H., & Chiang, M. T. (2010). Chitosan reduces gluconeogenesis and increases glucose uptake in skeletal muscle in streptozotocin-induced diabetic rats. *Journal of Agricultural and Food Chemistry*, *58*(9). https://doi.org/10.1021/jf100662r.

Liu, S.-H., Chen, R.-Y., & Chiang, M.-T. (2021). Effects and Mechanisms of Chitosan and ChitosanOligosaccharide on Hepatic Lipogenesis and Lipid Peroxidation, Adipose Lipolysis, and Intestinal Lipid Absorption in Rats with High-Fat Diet-Induced Obesity. *International Journal of Molecular Sciences*, *22*(3), 1139. https://doi.org/10.3390/ijms22031139.

Liu, Y., Shang, J., Chen, Y., & Feng, X. (2025). Potential Applications of Chitosan in Seborrheic Dermatitis and Other Skin Diseases: A Comprehensive Review. In *Clinical, Cosmetic and Investigational Dermatology* (Vol. 18, pp. 533–542). Dove Medical Press Ltd. https://doi.org/10.2147/CCID.S504778.

Lord, M. S., Cheng, B., McCarthy, S. J., Jung, M. S., & Whitelock, J. M. (2011). The modulation of platelet adhesion and activation by chitosan through plasma and extracellular matrix proteins. *Biomaterials*, *32*(28), 6655–6662. https://doi.org/10.1016/j.biomaterials.2011.05.062.

Lupascu, F. G., Sava, A., Tătărușanu, S.-M., Iacob, A.-T., Dascălu, A., Profire, B.-Ștefania, Vasincu, I.-M., Apotrosoaei, M., Gisca, T.-C., Turin-Moleavin, I.-A., & Profire, L. (2024). New Chitosan-Based Co-Delivery Nanosystem for Diabetes Mellitus Therapy. *Polymers*, *16*(13), 1825. https://doi.org/10.3390/polym16131825.

Macías, N., Vega, A., Abad, S., Aragoncillo, I., García-Prieto, A. M., Santos, A., Torres, E., & Luño, J. (2019). Middle molecule elimination in expanded haemodialysis: Only convective transport? *Clinical Kidney Journal*, *12*(3), 447–455. https://doi.org/10.1093/ckj/sfy097.

Marin, L., Andreica, B. I., Anisiei, A., Cibotaru, S., Bardosova, M., Materon, E. M., & Oliveira, O. N. (2023). Quaternized chitosan (nano)fibers: A journey from preparation to high performance applications. In *International Journal of Biological Macromolecules* (Vol. 242). Elsevier B.V. https://doi.org/10.1016/j.ijbiomac.2023.125136.

Martingo, M., Baptista-Silva, S., Mesquita, R., Ferreira, J. P., Borges, S., & Pintado, M. (2025). Exploring the potential of mealworm chitosan for hemodialysis applications. *Sustainable Chemistry and Pharmacy*, *45*. https://doi.org/10.1016/j.scp.2025.102013.

Martingo, M., Baptista-Silva, S., Pintado, M., & Borges, S. (2025). A new vision upon hemodialysis: A shift from synthetic to sustainable chitosan membranes. *Biomedical Engineering Advances*, *9*, 100175. https://doi.org/10.1016/j.bea.2025.100175.

Mascarenhas-Melo, F., Carvalho, A., Gonçalves, M. B. S., Paiva-Santos, A. C., & Veiga, F. (2022). Nanocarriers for the topical treatment of psoriasis - pathophysiology, conventional treatments, nanotechnology, regulatory and toxicology. In *European Journal of Pharmaceutics and Biopharmaceutics* (Vol. 176, pp. 95–107). Elsevier B. V. https://doi.org/10.1016/j.ejpb.2022.05.012.

Mesgari, M., Aalami, A. H., & Sahebkar, A. (2021). Antimicrobial activities of chitosan/titanium dioxide composites as a biological nanolayer for food preservation: A review. In *International Journal of Biological Macromolecules* (Vol. 176, pp. 530–539). Elsevier B. V. https://doi.org/10.1016/j.ijbiomac.2021.02.099.

Mirda, E., Idroes, R., Khairan, K., Tallei, T. E., Ramli, M., Earlia, N., Maulana, A., Idroes, G. M., Muslem, M., & Jalil, Z. (2021). Synthesis of silver nanoparticles–chitosan composite particles spheres and their antimicrobial activities. *Polymers*, *13*(22). https://doi.org/10.3390/polym13223990.

Mohandas, A., Anisha, B. S., Chennazhi, K. P., & Jayakumar, R. (2015). Chitosan–hyaluronic acid/VEGF loaded fibrin nanoparticles composite sponges for enhancing angiogenesis in wounds. *Colloids and Surfaces B: Biointerfaces*, *127*, 105–113. https://doi.org/https://doi.org/10.1016/j.colsurfb.2015.01.024.

Mohd Nordin, U. U., Ahmad, N., Salim, N., & Mohd Yusof, N. S. (2021). Lipid-based nanoparticles for psoriasis treatment: a review on conventional treatments, recent works, and future prospects. In *RSC Advances* (Vol. 11, Issue 46, pp. 29080–29101). Royal Society of Chemistry. https://doi.org/10.1039/d1ra06087b.

Monteiro, O. A. C., & Airoldi, C. (1999). Some studies of crosslinking chitosan-glutaraldehyde interaction in a homogeneous system. In *International Journal of Biological Macromolecules* (Vol. 26). www.elsevier.com/locate/ijbiomac.

Moran, H. B. T., Turley, J. L., Andersson, M., & Lavelle, E. C. (2018). Immunomodulatory properties of chitosan polymers. In *Biomaterials* (Vol. 184, pp. 1–9). Elsevier Ltd. https://doi.org/10.1016/j.biomaterials.2018.08.054.

Moraru, C., Mincea, M. M., Frandes, M., Timar, B., & Ostafe, V. (2018). A meta-analysis on randomised controlled clinical trials evaluating the effect of the dietary supplement chitosan on weight loss, lipid parameters and blood pressure. In *Medicina (Lithuania)* (Vol. 54, Issue 6). https://doi.org/10.3390/MEDICINA54060109.

Morisi, N., Alfano, G., Ferrarini, M., Ferri, C., Fontana, F., Ballestri, M., & Donati, G. (2025). Hemoadsorption and plasma adsorption: two current options for the 3rd dimension of dialysis purification. In *Journal of Nephrology* (Vol. 38, Issue 3, pp. 845–857). Springer Science and Business Media Deutschland GmbH. https://doi.org/10.1007/s40620-025-02257-x.

Msaad, R., Essadik, R., Mohtadi, K., Meftah, H., Lebrazi, H., Taki, H., Kettani, A., Madkouri, G., Ramdani, B., & Saïle, R. (2019). Predictors of mortality in hemodialysis patients. *Pan African Medical Journal*, *33*. https://doi.org/10.11604/pamj.2019.33.61.18083.

Mukherjee, S., Roy, D., & Bhattacharya, P. (2008). Comparative performance study of polyethersulfone and polysulfone membranes for trypsin isolation from goat pancreas using affinity ultrafiltration. *Separation and Purification Technology*, *60*(3), 345–351. https://doi.org/10.1016/j.seppur.2007.12.013.

Muñoz-Terol, J. M., Rocha, J. L., Castro-de la Nuez, P., Egea-Guerrero, J. J., Gil-Sacaluga, L., García-Cabrera, E., & Vilches-Arenas, A. (2023). Prognosis Factors of Patients Undergoing Renal Replacement Therapy. *Journal of Personalized Medicine*, *13*(4). https://doi.org/10.3390/jpm13040605.

Muxika, A., Etxabide, A., Uranga, J., Guerrero, P., & de la Caba, K. (2017). Chitosan as a bioactive polymer: Processing, properties and applications. *International Journal of Biological Macromolecules*, *105*, 1358–1368. https://doi.org/https://doi.org/10.1016/j.ijbiomac.2017.07.087.

Mydlík, M., Derzsiová, K., Rácz, O., Šipulová, A., Lovásová, E., Molčányiová, A., & Petrovičová, J. (2004). Vitamin E-coated dialyzer and antioxidant defense parameters: Three-month study. *Seminars in Nephrology*, *24*(5), 525–531. https://doi.org/https://doi.org/10.1016/j.semnephrol.2004.06.017.

Nasr, S., Rady, M., Gomaa, I., Syrovet, T., Simmet, T., Fayad, W., & Abdel-Kader, M. (2019). Ethosomes and lipid-coated chitosan nanocarriers for skin delivery of a chlorophyll derivative: A potential treatment of squamous cell carcinoma by photodynamic therapy. *International Journal of Pharmaceutics*, *568*. https://doi.org/10.1016/j.ijpharm.2019.118528.

Nguyen, H. T.-T., Tran, T. N., Ha, A. C., & Huynh, P. D. (2022). Impact of Deacetylation Degree on Properties of Chitosan for Formation of Electrosprayed Nanoparticles. *Journal of Nanotechnology*, *2022*(1), 2288892. https://doi.org/https://doi.org/10.1155/2022/2288892.

Noureen, S., Niaz, A., Zaman, I., & Tariq, M. (2025). Electrochemical Determination of Uric Acid in Biological Samples by Using Chitosan-Nickel(II) Complex Modified MWCNTs Paste Electrode. *Electrocatalysis*. https://doi.org/10.1007/s12678-025-00978-y.

Oh, S.-J., Nguyen, T. T., Seo, Y., Park, H.-J., Ahn, J.-S., Shin, Y. Y., Kang, B.-J., Jang, M., Park, J., Jeong, J.-H., & Kim, H.-S. (2024). Sustained release of stem cell secretome from nano-villi chitosan microspheres for effective treatment of atopic dermatitis. *International Journal of Biological Macromolecules*, *277*, 134344. https://doi.org/https://doi.org/10.1016/j.ijbiomac.2024.134344.

Pandya, D., Nagrajappa, A. K., & Ravi, K. S. (2016). Assessment and correlation of urea and creatinine levels in saliva and serum of patients with chronic kidney disease, diabetes and hypertension– A research study. *Journal of Clinical and Diagnostic Research*, *10*(10), ZC58–ZC62. https://doi.org/10.7860/JCDR/2016/20294.8651.

Patel, M. P., Patel, R. R., & Patel, J. K. (2010). Chitosan Mediated Targeted Drug Delivery System: A Review. In *J Pharm Pharmaceut Sci (www.cspsCanada.org)* (Vol. 13, Issue 3). www.cspsCanada.org.

Pereira, R. F., & Bártolo, P. J. (2014). Traditional Therapies for Skin Wound Healing. *Advances in Wound Care*, *5*(5), 208–229. https://doi.org/10.1089/wound.2013.0506.

Perzynski, L. A. (2025). Update on conventional therapy for atopic dermatitis. In *Allergo Journal International*. Springer Medizin. https://doi.org/10.1007/s40629-025-00346-8.

Prados-Carmona, A., Husein-ElAhmed, H., Navarro-Triviño, F. J., & Ruiz-Villaverde, R. (2025). Atopic Dermatitis beyond Eczema: A Review on its Systemic Impact through Pruritus and Associated Comorbidities. In *Clinical Reviews in Allergy and Immunology* (Vol. 68, Issue 1). Springer. https://doi.org/10.1007/s12016-025-09075-w.

Priyanka, D. N., Prashanth, K. V. H., & Tharanathan, R. N. (2022). A review on potential anti-diabetic mechanisms of chitosan and its derivatives. In *Carbohydrate Polymer Technologies and Applications* (Vol. 3). https://doi.org/10.1016/j.carpta.2022.100188.

Rabea, E., Badawy, M., Stevens, C., & Smagghe, G. (2003). Chitosan as Antimicrobial Agent: Applications and Mode of Action. *Biomacromolecules*, *4*, 1457–1465. https://doi.org/10.1021/bm034130m.

Radu, E. R., & Voicu, S. I. (2022). Functionalized Hemodialysis Polysulfone Membranes with Improved Hemocompatibility. In *Polymers* (Vol. 14, Issue 6). MDPI. https://doi.org/10.3390/polym14061130.

Rahnama, Shahrbanoo, Movaffagh, Jebrail, Shahroodi, Azadeh, Jirofti, Nafiseh, Fazly Bazzaz, Bibi Sedigheh, Beyraghdari, Monir, Hashemi, Maryam, & Kalalinia, Fatemeh. (2022). Development and characterization of the electrospun melittin-loaded chitosan nanofibers for treatment of acne vulgaris in animal model. *Journal of Industrial Textiles*, *52*, 15280837221112410. https://doi.org/10.1177/15280837221112410.

Reynolds, R. V., Yeung, H., Cheng, C. E., Cook-Bolden, F., Desai, S. R., Druby, K. M., Freeman, E. E., Keri, J. E., Stein Gold, L. F., Tan, J. K. L., Tollefson, M. M., Weiss, J. S., Wu, P. A., Zaenglein, A. L., Han, J. M., & Barbieri, J. S. (2024). Guidelines of care for the management of acne vulgaris. *Journal of the American Academy of Dermatology*, *90*(5), 1006.e1-1006.e30. https://doi.org/10.1016/j.jaad.2023.12.017.

Roky, A. H., Islam, M. M., Ahasan, A. M. F., Mostaq, M. S., Mahmud, M. Z., Amin, M. N., & Mahmud, M. A. (2025). Overview of skin cancer types and prevalence rates across continents. In *Cancer Pathogenesis and Therapy* (Vol. 3, Issue 2, pp. 89–100). Chinese Medical Association. https://doi.org/10.1016/j.cpt.2024.08.002.

Saeedi, M., Morteza-Semnani, K., Akbari, J., Hajheydari, Z., Goodarzi, A., Rostamkalaei, S. S., Hassan Hashemi, S. M., & Rahimnia, S. M. (2024). Green Formulation of Spironolactone Loaded Chitosan-Coated Nano Lipid Carrier for Treatment of Acne Vulgaris: A Randomized Double-Blind Clinical Trial. *Advanced Pharmaceutical Bulletin*, *14*(1), 161–175. https://doi.org/10.34172/apb.2024.011.

Salehi Moghaddam, A., Bahrami, M., Sarikhani, E., Tutar, R., Ertas, Y. N., Tamimi, F., Hedayatnia, A., Jugie, C., Savoji, H., Qureshi, A. T., Rizwan, M., Maduka, C. V., & Ashammakhi, N. (2025). Engineering the Immune Response to Biomaterials. In *Advanced Science* (Vol. 12, Issue 19). John Wiley and Sons Inc. https://doi.org/10.1002/advs.202414724.

Sarkar, S., Das, D., Dutta, P., Kalita, J., Wann, S. B., & Manna, P. (2020). Chitosan: A promising therapeutic agent and effective drug delivery system in managing diabetes mellitus. In *Carbohydrate Polymers* (Vol. 247). https://doi.org/10.1016/j.carbpol.2020.116594.

Scarpioni, R., Ricardi, M., Albertazzi, V., De Amicis, S., Rastelli, F., & Zerbini, L. (2016). Dialysis-related amyloidosis: Challenges and solutions. In *International Journal of Nephrology and Renovascular Disease* (Vol. 9, pp. 319–328). Dove Medical Press Ltd. https://doi.org/10.2147/IJNRD.S84784.

Shafiee, S., Dastmalchi, S., Gharekhani, A., & Shayanfar, A. (2024). Analysis of indoxyl sulfate in biological fluids with emphasis on sample preparation techniques: A comprehensive analytical review. In *Heliyon* (Vol. 10, Issue 15). Elsevier Ltd. https://doi.org/10.1016/j.heliyon.2024.e35032.

Shannon, C. M., Ballew, S. H., Daya, N., Zhou, L., Chang, A. R., Sang, Y., Coresh, J., Selvin, E., & Grams, M. E. (2021). Serum albumin and risks of hospitalization and death: Findings from the Atherosclerosis Risk in Communities study. *Journal of the American Geriatrics Society*, *69*(10), 2865–2876. https://doi.org/10.1111/jgs.17313.

Sheikh, A., Hazari, S. A., Molugulu, N., Alshehri, S. A., Wahab, S., Sahebkar, A., & Kesharwani, P. (2023). Hyaluronic acid engineered gallic acid embedded chitosan nanoparticle as an effective delivery system for treatment of psoriasis. *Environmental Research*, *238*, 117086. https://doi.org/https://doi.org/10.1016/j.envres.2023.117086.

Shi, S.-C., & Chang, Y.-W. (2022). Biofriendly chitosan-based high-efficiency dialysis membrane. *Progress in Organic Coatings*, *170*, 106981. https://doi.org/https://doi.org/10.1016/j.porgcoat.2022.106981.

Sirisuksakun, C., Satirapoj, B., Thimachai, P., & Supasyndh, O. (2025). Amelioration of uremic toxin indoxyl sulfate by oral chito-oligosaccharide in predialysis patients. In *Journal of Nephrology*. Springer Science and Business Media Deutschland GmbH. https://doi.org/10.1007/s40620-025-02279-5.

Skinner, S. C., Derebail, V. K., Poulton, C. J., Bunch, D. C., Roy-Chaudhury, P., & Key, N. S. (2021). Hemodialysis-Related Complement and Contact Pathway Activation and Cardiovascular Risk: A Narrative Review. In *Kidney Medicine* (Vol. 3, Issue 4, pp. 607–618). Elsevier Inc. https://doi.org/10.1016/j.xkme.2021.04.006.

Soeters, P. B., Wolfe, R. R., & Shenkin, A. (2019). Hypoalbuminemia: Pathogenesis and Clinical Significance. In *Journal of Parenteral and Enteral Nutrition* (Vol. 43, Issue 2, pp. 181–193). John Wiley and Sons Inc. https://doi.org/10.1002/jpen.1451.

Soglas, I., Khutoryanskiy, V., & Williams, A. (2010). Exploring the Factors Affecting the Solubility of Chitosan in Water. *Macromolecular Chemistry and Physics*, *211*, 426–433. https://doi.org/10.1002/macp.200900385.

Suryani, S., Chaerunisaa, A. Y., Joni, I. M., Ruslin, R., Ramadhan, L. O. A. N., Wardhana, Y. W., & Sabarwati, S. H. (2022). Production of Low Molecular Weight Chitosan Using a Combination of Weak Acid and Ultrasonication Methods. *Polymers*, *14*(16). https://doi.org/10.3390/polym14163417.

Tao, W., Wang, G., & Wei, J. (2021). The role of chitosan oligosaccharide in metabolic syndrome: A review of possible mechanisms. In *Marine Drugs* (Vol. 19, Issue 9). https://doi.org/10.3390/md19090501.

Tarrass, F., & Benjelloun, M. (2022). Health and environmental effects of the use of N-methyl-2-pyrrolidone as a solvent in the manufacture of hemodialysis membranes: A sustainable reflexion. *Nefrologia*, *42*(2), 122–124. https://doi.org/10.1016/j.nefro.2021.05.003.

Teuwafeu, D. G., Sehbing, M., Halle, M. P., Mahamat, M., Fouda, H., & Ashuntantang, G. (2024). Quality of life and social reinsertion of patients on maintenance haemodialysis in four government funded hospitals in Cameroon. *BMC Nephrology*, *25*(1). https://doi.org/10.1186/s12882-024-03778-6.

Tzeng, H. P., Liu, S. H., & Chiang, M. T. (2022). Antidiabetic Properties of Chitosan and Its Derivatives. In *Marine Drugs* (Vol. 20, Issue 12). https://doi.org/10.3390/md20120784.

Valades-Aguilar, B. A., Rivera-González, T. I., Rangel-López, R., Luna-Barcenas, G., Franco-Molina, M. Á., Rodriguez-Padilla, C., & Zárate-Triviño, D. G. (2024). Influence of Chitosan Purification on the Immunomodulator Potential of Chitosan Nanoparticles on Human Monocytes. *Polymers*, *16*(23). https://doi.org/10.3390/polym16233390.

Vanholder, R., Schepers, E., Pletinck, A., Nagler, E. V, & Glorieux, G. (2014). The Uremic Toxicity of Indoxyl Sulfate and p-Cresyl Sulfate: A Systematic Review. *Journal of the American Society of Nephrology*, *25*(9). https://journals.lww.com/jasn/fulltext/2014/09000/the_uremic_toxicity_of_indoxyl_sulfate_and.7.aspx.

Vasam, M., Korutla, S., & Bohara, R. A. (2023). Acne vulgaris: A review of the pathophysiology, treatment, and recent nanotechnology based advances. In *Biochemistry and Biophysics Reports* (Vol. 36). Elsevier B.V. https://doi.org/10.1016/j.bbrep.2023.101578.

Vega-Vega, O., Ronco, C., & Martínez-Rueda, A. J. (2023). Pushing the boundaries of hemodialysis: innovations in membranes and sorbents. In *Revista de investigacion clinica; organo del Hospital de Enfermedades de la Nutricion* (Vol. 75, Issue 6, pp. 274–288). https://doi.org/10.24875/RIC.23000223.

Verma, C., & Quraishi, M. A. (2021). Chelation capability of chitosan and chitosan derivatives: Recent developments in sustainable corrosion inhibition and metal decontamination applications. In *Current Research in Green and Sustainable Chemistry* (Vol. 4). Elsevier B.V. https://doi.org/10.1016/j.crgsc.2021.100184.

Viala, S., Freche, " M, & Lacout, J. L. (1996). Effect of chitosan on octacalcium phosphate crystal growth. In *Carhohydratr Polymers* (Vol. 29).

Victor, R. de S., Santos, A. M. da C., de Sousa, B. V., Neves, G. de A., Santana, L. N. de L., & Menezes, R. R. (2020). A review on Chitosan's uses as biomaterial: Tissue engineering, drug delivery systems and cancer treatment. In *Materials* (Vol. 13, Issue 21, pp. 1–71). MDPI AG. https://doi.org/10.3390/ma13214995.

Wang, Z., Yan, Y., Zhang, Z., Li, C., Mei, L., Hou, R., Liu, X., & Jiang, H. (2024). Effect of Chitosan and Its Water-Soluble Derivatives on Antioxidant Activity. *Polymers*, *16*(7). https://doi.org/10.3390/polym16070867.

Wolley, M., Jardine, M., & Hutchison, C. A. (2018). Exploring the Clinical Relevance of Providing Increased Removal of Large Middle Molecules. *Clinical Journal of the American Society of Nephrology*, *13*(5). https://journals.lww.com/cjasn/fulltext/2018/05000/exploring_the_clinical_relevance_of_providing.25.aspx.

Wu, F.-C., Tseng, R.-L., & Juang, R.-S. (1999). Role of pH in Metal Adsorption from Aqueous Solutions Containing Chelating Agents on Chitosan. *Industrial & Engineering Chemistry Research*, *38*(1), 270–275. https://doi.org/10.1021/ie980242w.

Xiong, S., Lyu, Y., Davenport, A., & Choy, K. L. (2021). Sponge-like chitosan based porous monolith for uraemic toxins sorption. *Nanomaterials*, *11*(9). https://doi.org/10.3390/nano11092247.

Yadav, H., Malviya, R., & Kaushik, N. (2024). Chitosan in biomedicine: A comprehensive review of recent developments. *Carbohydrate Polymer Technologies and Applications*, *8*. https://doi.org/10.1016/j.carpta.2024.100551.

Yamamoto, S. (2019). Molecular mechanisms underlying uremic toxin-related systemic disorders in chronic kidney disease: focused on β 2 -microglobulin-related amyloidosis and indoxyl sulfate-induced atherosclerosis—Oshima Award Address 2016. In *Clinical and Experimental Nephrology* (Vol. 23, Issue 2, pp. 151–157). Springer Tokyo. https://doi.org/10.1007/s10157-018-1588-9.

Yan, S., & Qiu, Y. (2024). Improving Hemocompatibility of Polysulfone Membrane by UV-Assisted Grafting of Sulfonated Chitosan. *Polymers*, *16*(11). https://doi.org/10.3390/polym16111555.

Yang, S., Diao, Z., Liu, W., & Guo, W. (2023). Effect of dialytic phosphate reduction rate on mortality in maintenance hemodialysis patients: a matched case–control study. *BMC Nephrology*, *24*(1). https://doi.org/10.1186/s12882-023-03199-x.

Yeo, G., & Kim, J. (2025). Antibacterial Chitosan-Based Double-Network Hydrogel Patch Loaded with Antioxidant Ceria Nanoparticles and Betamethasone to Treat Psoriasis. *Biomacromolecules*, *26*(3), 1627–1638. https://doi.org/10.1021/acs.biomac.4c01442.

Yuan, W. P., Liu, B., Liu, C. H., Wang, X. J., Zhang, M. S., Meng, X. M., & Xia, X. K. (2009). Antioxidant activity of chito-oligosaccharides on pancreatic islet cells in streptozotocin-induced diabetes in rats. *World Journal of Gastroenterology*, *15*(11). https://doi.org/10.3748/wjg.15.1339.

Zawada, A. M., Lang, T., Ottillinger, B., Kircelli, F., Stauss-Grabo, M., & Kennedy, J. P. (2022). Impact of Hydrophilic Modification of Synthetic Dialysis Membranes on Hemocompatibility and Performance. In *Membranes* (Vol. 12, Issue 10). MDPI. https://doi.org/10.3390/membranes12100932.

Zeng, L., Gowda, B. H. J., Ahmed, M. G., Abourehab, M. A. S., Chen, Z. S., Zhang, C., Li, J., & Kesharwani, P. (2023). Advancements in nanoparticle-based treatment approaches for skin cancer therapy. In *Molecular Cancer* (Vol. 22, Issue 1). BioMed Central Ltd. https://doi.org/10.1186/s12943-022-01708-4.

Zhang, H., Zhang, Y., Li, L., Zhao, S., Ni, H., Cao, S., & Wang, J. (2014). Cross-linked polyacrylonitrile/polyethyleneimine-polydimethylsiloxane composite membrane for solvent resistant nanofiltration. *Chemical Engineering Science*, *106*, 157–166. https://doi.org/10.1016/j.ces.2013.11.043.

Zhang, J., Guo, S., Yang, H., Yang, L., & Tan, X. (2025). Interfacial polymerization of poly(ethylenimine) on PAN hollow fibers for direct air capture of CO2. *Journal of Environmental Chemical Engineering*, *13*(2). https://doi.org/10.1016/j.jece.2025.116077.

Zheng, H., Li, Y., Li, W., Zhou, S., Huang, C., & Du, L. (2024). Ultrasound induced biosynthesis of silver nanoparticles embedded into chitosan polymers: Investigation of its anti-cutaneous squamous cell carcinoma effects. *Open Chemistry*, *22*(1). https://doi.org/10.1515/chem-2024-0018.

Zhu, B., Jing, M., Yu, Q., Ge, X., Yuan, F., & Shi, L. (2022). Treatments in psoriasis: from standard pharmacotherapy to nanotechnology therapy. *Advances in Dermatology and Allergology/Postępy Dermatologii i Alergologii*, *39*(3), 460–471. https://doi.org/10.5114/ada.2021.108445.

Zosangpuii, & Sudheer, P. (2024). Chitosan a natural anti-inflammatory and wound healing agent: A brief update. In *Natural Resources for Human Health* (Vol. 4, Issue 2, pp. 115–127). Visagaa Publishing House. https://doi.org/10.53365/nrfhh/177602.

Chapter 3

Chitosan: A Multifunctional Biological Polymer

Li Huang[1], PhD
Wenyan Liu[1]
Hui Zhang[1], MA
Song Jin[2]
Wei Wang[1]
Xianzhao Shao[1]
and Xiang Nan[1,*], PhD
[1]Shaanxi Key Laboratory of Catalysis, School of Chemical and Environmental Science, Shaanxi University of Technology, Hanzhong, P. R. China
[2]Faculty of Chemistry, Biology and Geography,
Kyrgyz National University named after Ishenaly Arabaev, Bishkek, Kyrgyzstan

Abstract

Chitosan is a natural linear polysaccharide functional material, formed by the β-(1, 4)-glycosidic bond between D-glucosamine and N-acetyl-D-glucosamine. The free amino groups present within its molecular structure confer distinctive physicochemical characteristics, including excellent biocompatibility, degradability, antibacterial activity, and metal ion chelation capability. Chitosan is derived from diverse biological sources, encompassing marine crustacean waste, fungal cell walls, and insect exoskeletal structures. However, traditional preparation methods show problems such as significant molecular weight loss and environmental pollution, prompting researchers to explore new preparation techniques. Over the past few years, a suite of modern green

* Corresponding Author's Email: nanxiangniu@163.com

In: Chitosan
Editor: Denis Hurst
ISBN: 979-8-90134-079-0

preparation technologies, such as enzymatic deacetylation, microbial pretreatment, and physical field enhancement approaches, have been increasingly investigated and characterized. These technologies are known for their mildness, efficiency, and environmental friendliness, making it possible to produce chitosan. Additionally, quality control is the key to ensuring the stable and safe performance of chitosan during the preparation process, including raw material selection, process monitoring, and product characterization. By establishing a full chain traceability system, intelligent reaction control system, and multi-dimensional detection system, the stability and reliability of chitosan product quality are ensured. This chapter comprehensively reviews the sources, preparation processes, and quality control of chitosan, and deeply analyzes the effects of different sources and processes on the properties of chitosan. This further reveals the technological bottlenecks and development trends faced by chitosan, providing theoretical support and technical guidance for the further development of the chitosan industry.

Keywords: chitosan, source, biocompatibility, preparation, quality control, sustainable production

1. Introduction

Chitosan is a natural polysaccharide whose history can be traced back to the middle of the 19th century. In 1859, Rouget first synthesized chitosan by treating chitosan with concentrated alkali, revealing an important biological molecule (Aberoumand and Hoseinian, 2025; da Silva Ponciano et al., 2025). The molecular skeleton of chitosan is composed of alternating D-glucosamine residues linked by β-(1, 4) glycosidic bonds and N-acetyl-D-glucosamine residues (Figure 1). The unsubstituted amino groups in its structure endow chitosan with unique physicochemical properties (Basso et al., 2025; Deshmukh et al., 2025). Specifically, it involves good tissue compatibility, controllable biodegradability, natural antibacterial efficacy, self-assembly into films, and complexation ability with metal ions. With these distinctive properties, chitosan has become a cornerstone of modern scientific exploration, securing its leading role in interdisciplinary studies covering biomaterials, pharmaceuticals, ecological conservation, and sustainable agriculture. (El-Naggar et al., 2025). Due to its remarkable biocompatibility and biodegradability in medicine, chitosan has been widely recognized as an optimal candidate for developing advanced drug delivery systems and three-dimensional tissue engineering scaffolds. In terms of environmental protection, chitosan demonstrates the ability to chelate metal ions, making it important in wastewater treatment (Chevrier et al., 2025). In agriculture, the antibacterial and film-forming properties of chitosan hold promising application prospects for plant protection and seed coating. (Basso et al., 2025). With the continuous advancement of science and technology, people are constantly exploring the uniqueness and broad application prospects of chitosan.

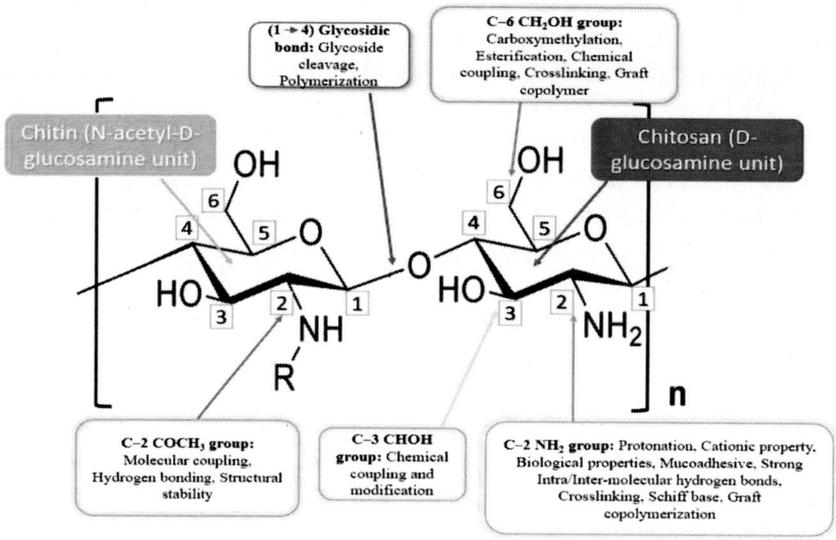

Figure 1. Molecular structure of chitosan [Reproduced with permission from Elsevier, Ref. (Olunusi et al., 2025)].

Chitosan is derived from a wide range of raw materials, with marine and freshwater crustaceans being one of the most traditional raw materials (Drebezghova et al., 2025). Every year, a large amount of crustacean processing by-products, such as shrimp shells and crab shells, are chemically extracted to produce high-purity chitosan. However, traditional preparation protocols are associated with notable deficiencies like molecular weight reduction and environmental hazards, motivating researchers to pursue innovative green fabrication technologies. Recently, modern green preparation technologies such as enzyme deacetylation, microbial pretreatment and physical field enhancement have made significant progress (Han et al., 2025). Quality control represents a pivotal step in ensuring the stability and safety of the chitosan preparation process. From raw material selection, process monitoring, product characterization to the formulation of standard system, each step directly affects the final quality of chitosan. For example, the quality of raw chitosan directly affects the purity and performance of chitosan. It is necessary to accurately control the deacetylation degree, molecular weight and other parameters to ensure the consistency of product quality (Ibrahim et al., 2025).

Despite significant advancements in both the fundamental studies and translational applications of chitosan, several critical challenges remain to be addressed. Technically, precise control of deacetylation level and molecular weight is a key scientific challenge. While the conventional concentrated alkali approach can yield chitosan with a high degree of deacetylation, it is accompanied by substantial molecular weight degradation. Enzymatic deacetylation is facing the technical bottleneck of high enzyme cost and upper deacetylation degree while maintaining molecular weight. Additionally, the stability and yield of microbial fermentation processes need improvement to meet industrial production demands. In environmental sustainability, the high pollution and energy consumption problems of traditional methods have not been fundamentally resolved (Huang et al., 2025). Therefore, it is necessary to develop green preparation technologies with zero emissions and low energy consumption. With respect to market entry and standardization frameworks, the absence of harmonized global quality standards for chitosan, together with stringent technical trade barriers, constitutes a critical bottleneck impeding the globalization of the chitosan sector. To tackle these critical challenges, this chapter presents a systematic and up-to-date review of the origin, advanced preparation technologies, and rigorous quality control strategies for chitosan. The effects of different sources and processes on the properties of chitosan are analyzed, and the technical bottlenecks and development trends faced by the industry are illustrated (Long et al., 2025b). By proposing targeted solutions and strategies, this chapter aims to provide theoretical support and technical guidance for the further development of the chitosan industry, promoting its transition towards a greener, more efficient, and sustainable direction.

2. Sources and Origin of Chitosan

Chitosan is a natural polysaccharide functional material. Driven by the broad diversity of its raw materials and ongoing innovations in fabrication technologies, chitosan has gained notable industrial applicability. In the traditional industry, marine and freshwater crustaceans are still a major source of chitosan (Figure 2). Eight million tons of by-products from aquatic product processing worldwide are converted into high-purity chitosan through chemical extraction processes (Kumar et al., 2025; Li et al., 2019a). Meanwhile, fungal cell walls are emerging with unique molecular structure characteristics and biological safety advantages. *Aspergillus* strains can customize the synthesis of natural chitosan with a deacetylation degree of 70% through fermentation processes, and their pharmaceutical-grade products show significant potential in tissue engineering (Lin et al., 2025). More importantly, breakthroughs in insect epidermis and synthetic biology have opened up new areas for the chitosan industry. The thermal stability of fly maggot chitosan is improved compared with traditional products. Moreover, gene-edited microorganisms can precisely biosynthesize customized chitosan from glucose. This multi-source technology is driving the advancement of chitosan preparation processes toward environmental sustainability and intelligent manufacturing, thereby establishing an innovative paradigm for the sustainable development of the biomaterials industry.

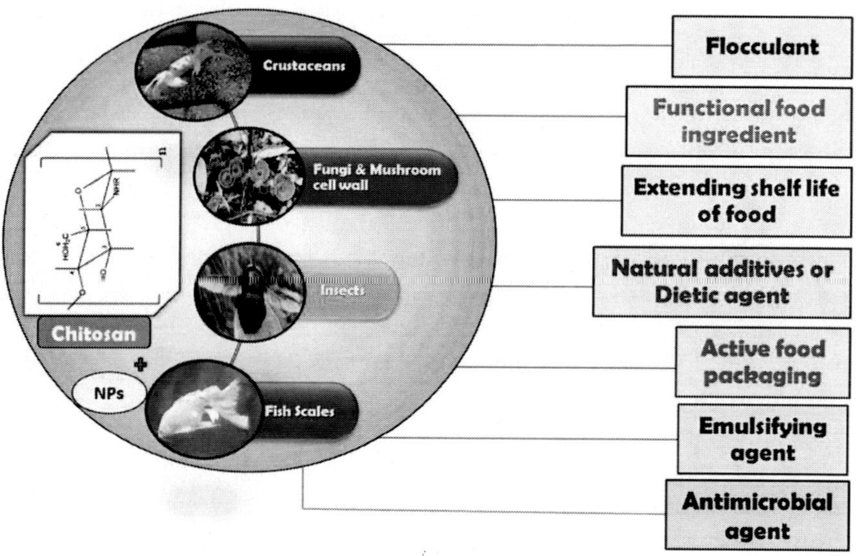

Figure 2. The source of chitosan [Reproduced with permission from Elsevier, Ref. (Mawale et al., 2025)].

2.1. Marine and Freshwater Crustacean Waste

The industrial production of chitosan relies heavily on the processing waste of marine and freshwater crustaceans, forming a global resource recycling network. The shells of aquatic products such as shrimp and crab are the core raw materials, and the content of chitosan varies by species. For example, the shell of Pacific white shrimp contains 15% to 20% chitosan, while the shell of Alaska king crab contains up to 40% chitosan (Martingo et al., 2025; Moussa et al., 2019). This change is mainly due to the different species of shell mineralization degree and protein crosslinking structure difference. Approximately 8 million tons of crustacean waste are generated globally each year. Notably, over 60% of these sources are concentrated in key aquatic product processing hubs across Asia, encompassing China, India, and various Southeast Asian countries.

The complex layered structure of crustaceans poses significant challenges to the extraction process. The shell is composed of the outer cortex, outer stratum corneum and inner stratum corneum, and each layer forms a three-dimensional network of chitosan-protein-calcium carbonate complexes (Lieu et al., 2025). Chitosan is generally obtained from the shells of crustaceans through traditional chemical methods in the following. Dilute hydrochloric acid is used to remove calcium carbonate, with precise control over acid concentration and reaction time to prevent chitosan degradation; then, strong alkali under high-temperature conditions is used to remove proteins (Mohseni et al., 2025; Nabil et al., 2025). However, this approach tends to induce the cleavage of chitosan molecular chains, thereby disrupting the molecular weight distribution of the final product. Ultimately, the oxidation and decolorization process may introduce environmental pollutants. Despite the energy consumption and pollution issues, each ton of waste can yield approximately 150-200 kg of high-purity chitosan, and the balance between economic value and environmental benefits is driving continuous process optimization (Putro et al., 2025).

The sustainable development potential of this source path is continuously being unleashed through technological innovation. The new enzymatic protein hydrolysis technology has entered the pilot production stage. Hydrolyzing proteins with specific proteases under mild conditions can reduce the use of chemical reagents by 80% (Rony et al., 2025). Membrane separation technology has been applied in the desulfurization and decolorization processes, raising the wastewater recovery rate to over 95%. (Rani et al., 2025). More importantly, some companies are exploring comprehensive utilization models, converting the extracted residues into feed additives or bio-fertilizers to achieve 100% resource utilization of crustacoid waste (Panda et al., 2025). This closed-loop production system not only reduces chitosan production costs but also transforms environmental burdens into economic benefits, setting a model for a circular economy in the global aquatic processing industry.

2.2. Fungal Cell Wall

Fungal chitosan is reshaping the industry model with its unique advantages. During liquid deep fermentation, mycelia of filamentous fungi such as *Aspergillus* and *Rhizobium* can accumulate pure chitosan or chitosan complexes accounting for 5% to 20% of their dry weight (Oliveira Júnior et al., 2025; Peng et al., 2025). Compared to animal sources, the structure of fungal cell walls is markedly different: chitosan is embedded in micro-fibers within the *β*-glucan matrix, forming a more regular network structure (Popyrina et al., 2025). This structural characteristic leads to a narrower molecular weight distribution and higher crystallinity in fungal chitosan. Above all, some fungi naturally undergo deacetylation during growth, allowing them to directly synthesize natural chitosan with a deacetylation degree exceeding 70%, thus eliminating the need for additional chemical processing steps.

The precise control of the fermentation process has opened up new avenues for the customized production of fungal chitosan. By optimizing the culture medium composition, fermentation parameters (pH 5.0–6.5, temperature 28–32°C), and dissolved oxygen levels, precise control over the molecular weight and degree of deacetylation of the resultant products can be achieved (Dejene and Abtew, 2025; El-Naggar et al., 2025). For instance, in batch fermentation, the addition of chitosan synthase inhibitors at different stages can induce the production of low-molecular-weight chitosan (50,000-100,000 daltons). Solid-state fermentation technology uses agricultural waste (such as rice husks and bran) as substrates, which not only reduces production costs but also endows chitosan with unique biological activity. Some strains have been shown to produce chitosan with significant antioxidant and immune-regulating properties.

The biosafety advantage of fungal chitosan is becoming the core competitive advantage in the high-end market. This product does not contain animal allergens, heavy metals, polychlorinated biphenyls and other marine pollutants. Its biocompatibility is significantly better than that of traditional products (Dong et al., 2019). Within the pharmaceutical and biomedical sectors, fungal chitosan serves as a key material for fabricating absorbable surgical sutures and tissue engineering scaffolds, and its degradation metabolite glucosamine holds significant potential for facilitating cartilage regeneration. In the cosmetics industry, chitosan showed excellent mildness and film-forming properties, which led to the development of anti-aging facial mask and sunscreen products (Ghorai et al., 2025). With the wide use of GMP level fermentation workshops, fungal chitosan is transitioning from special raw materials to bulk products. It is expected that its market share in the field of medicine and health will grow at an average annual rate of 15% in the next five years.

2.3. Insect Exoskeleton and Alternative Resources

The value of insect exoskeletons as a source of chitosan is being redefined. The global insect farming industry generates over 2 million tons of processing by-products each year, among which the shells of black fly larvae and silkworm pupae are rich in chitosan resources. Insect chitosan has a unique structure (Franconetti et al., 2019). Its molecular chains form dense crystalline regions through stronger hydrogen bonds, requiring higher energy input for deacetylation reactions (Gou et al., 2025; He et al., 2025). This structural characteristic endows insect-derived chitosan with excellent thermal stability and mechanical properties, thereby demonstrating promising application potential in food packaging films. Experimental data show that the tensile strength of black soldier fly chitosan film can reach 85 MPa, significantly surpassing the 60 MPa of shrimp-derived products, while maintaining good oxygen barrier properties (Khani et al., 2025).

Marine resources, such as squid cartilage, provide possibilities for the development of special chitosan. The β-chitosan found in squid rings has a unique parallel chain conformation. Compared with the α-chitosan presented in shrimp and crab, its solubility is more than three times higher, and it can be directly dissolved in dilute acetic acid (Manohar et al., 2025; Naveedunissa et al., 2025; Patel and Uppaluri, 2025). This characteristic gives β-chitosan a unique advantage in wound dressings. Its hydrogel formula can form a more uniform fiber network, promoting cell adhesion and proliferation (Piboonkunsamlit et al., 2025).

Synthetic biology is revolutionizing the traditional production methods of chitosan. The research team utilized metabolic engineering technology to modify *Escherichia coli* and successfully achieved the biosynthesis of chitosan from glucose, with a yield exceeding 5g/L (Hill and Tabor, 2025; Milbreta et al., 2025; Putra et al., 2025) Most remarkably, through the targeted introduction of the chitosan deacetylase gene into engineered yeast expression systems, it has become possible to directly synthesize chitosan with a precisely controlled degree of deacetylation in industrial-scale fermentation tanks. Besides no strong acid or alkali treatment is required. This "cell factory" model not only shortens the production cycle from 7 d to 48 h, but also enables precise design of product structures through gene editing (Lambrianidi et al., 2019). For instance, by using CRISPR technology to knock out specific domains of chitosan synthase, uniform polymers containing only N-acetylglucosamine units can be obtained, which exhibit excellent transfection efficiency in gene delivery applications (Li et al., 2025).

2.4. Artificial Preparation

The deacetylation process is a key technical challenge in chitosan production. The deacetylation degree of traditional concentrated alkali method can reach 70%-90%, but the molecular weight loss is obvious, and the polydispersity index (PDI) is higher than 3.0 (Lemine et al., 2025). The new microwave-assisted process, through selective heating, achieves efficient deacetylation at 80°C, reducing the reaction time from 8 h to just 2 h (Lu et al., 2025). Eco-friendly ionic liquids, such as 1-butyl-3-methylimidazolium chloride, can simultaneously dissolve and deacetylate chitosan under mild conditions, achieving a product molecular weight retention rate of over 85% and an ionic liquid recovery rate of more than 90%. (Liu et al., 2025).

The precise control of deacetylation degree (DDA) directly influences the functional properties of chitosan. When DDA exceeds 70%, chitosan becomes water-soluble (pH <6.5) due to the disruption of intramolecular hydrogen bonds caused by the protonation of amino groups (Liu et al., 2019). In the medicine, chitosan with 85% DDA is used as an adhesion enhancer for mucosal surfaces, as its amino density matches the binding sites on cell membrane glycoproteins. Chitosan with 60% DDA retains some acetyl groups, which enhances cell adhesion in tissue engineering. The molecular weight distribution is equally crucial to the structural and functional properties of chitosan (Long et al., 2025a). As low molecular weight (less than 50,000 daltons) chitosan oligosaccharides exhibit significant immunomodulatory activity, while high molecular weight (>500,000) products are better suited for use as film materials.

In the future, the preparation methods of chitosan will be developed towards intelligence and customization. The reaction condition prediction model based on machine learning achieve precise control of deacetylation degree within ±2%, and the process parameters are adjusted in real time through the online monitoring system (near-infrared spectroscopy, raman spectroscopy) (Rajeswari et al., 2025; Tan et al., 2025). More forward-looking studies have focused on the design of continuous flow reactors, transforming traditional mass production into continuous manufacturing and reducing energy consumption per unit of production by 40% (Shi et al., 2025). In terms of sustainability, breakthroughs have been made in bio-based alkaline solutions and electrochemical deacetylation technologies. The use of renewable energy to drive electrolytic reactions has significantly reduced the consumption of chemical reagents. These innovations will shift the chitosan industry from resource-dependent to technology-driven, ushering in a new era of biomaterial manufacturing.

3. Preparation of Chitosan

3.1. Traditional Preparation Methods

3.1.1. Raw Material Pretreatment: Chemical Extraction of Chitosan

For the industrial pretreatment of crustacean raw materials such as shrimp and crab, multi-step chemical treatment is adopted to obtain high-purity chitosan. Firstly, a 1-2 M HCl solution is used for descaling at a temperature of 60-100°C for 1 to 24 h (Salama and Hesemann, 2025). This acid hydrolysis reaction can dissolve 30–50% of the calcium carbonate in the shell, producing soluble calcium chloride and releasing carbon dioxide (Ye et al., 2025). This process generates a large amount of acidic wastewater containing calcium salts. Next, the material is subjected to protein removal by placing it in a 2-10% sodium hydroxide solution and treating it at high temperatures (65-100°C) for 1-48 h (Zeraat pisheh et al., 2025). The strong alkaline hydrolysis breaks the disulfide bonds, hydrogen bonds, and hydrophobic interactions between chitosan and proteins, generating highly alkaline wastewater rich in peptides. Eventually, decolorization and bleaching are performed using 1-5%

hydrogen peroxide at 40-80°C to oxidize and degrade carotenoid pigments like astaxanthin, improving the product's color (Zacheski et al., 2025). The entire pretreatment process consumes a significant amount of acid and alkali reagents, resulting in complex pollution wastewater with high salinity and high chemical oxygen demand (COD). Additionally, the prolonged high-temperature environment causes partial depolymerization of the chitosan molecular chains, leading to a 10-30% loss of molecular weight and posing quality risks for subsequent processes.

3.1.2. Deacetylation by Concentrated Alkali and High Temperature: Core Chemical Transformation

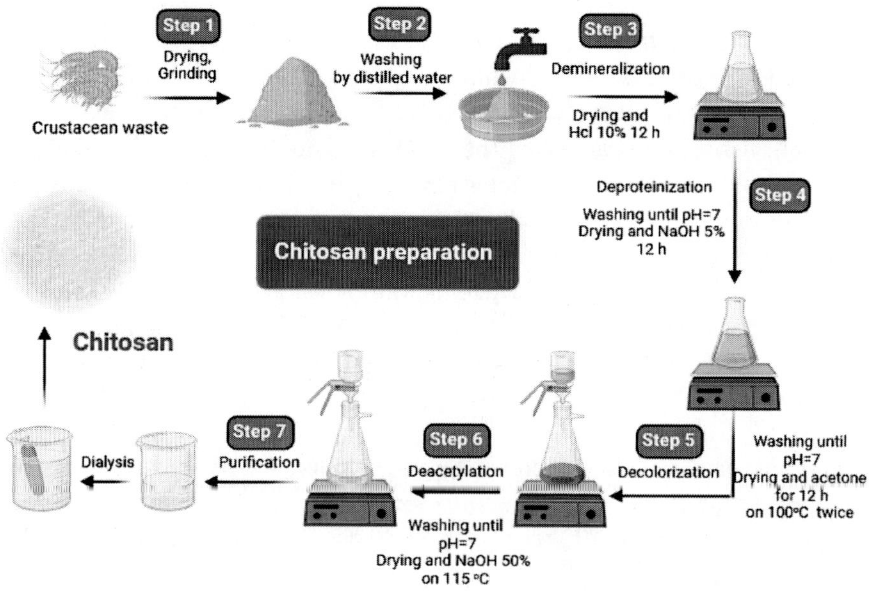

Figure 3. An innovative way for producing chitosan from marine animal waste [Reproduced with permission from Elsevier, Ref. (El-Saadony et al., 2025)].

Deacetylation, a core step in traditional ways, is carried out in a 40-50% sodium hydroxide solution (Figure 3). The chitosan powder after pretreatment undergoes high-temperature treatment of 80-140°C for 3-10 h in strong alkaline medium (Vieira et al., 2025). Under alkaline catalysis, the acetyl groups ($-NHCOCH_3$) attached to the N-acetylglucosamine units of the molecular chains undergo hydrolysis, releasing acetate ions and converting into reactive free amino groups ($-NH_2$), thereby forming the structure of chitosan. Although these intense reaction conditions can increase the deacetylation degree (DD) of chitosan to over 95%, the high-temperature alkaline environment can also lead to extensive cleavage of *β*-(1, 4) - glycosidic bonds on the main chain of chitosan (Silva et al., 2025; Tuesta et al., 2025). This effect results in a significant decline in molecular weight, dropping from the range of 100–1000 kDa to 5–50 kDa, which corresponds to a molecular weight loss rate of over 70%. To mitigate molecular chain degradation, 0.5-2% sodium borohydride ($NaBH_4$) is commonly added as a protective agent in industry. This approach reduces molecular weight loss by approximately 15% through the protective effect of its reducing end groups, but it still cannot fundamentally solve the problem of a significant decrease in the degree of polymerization of the products.

3.1.3. Product Refinement: Post-Processing and Quality Defects

The crude product after deacetylation needs to be systematically purified. Primarily, the product is rinsed thoroughly with copious amounts of deionized water until neutrality is achieved (pH ≈ 7), so as to eliminate residual alkali and soluble salts. Next, the product is dissolved in a 1% acetic acid solution to form a viscous gel, and then filtered and centrifuged to remove unreacted chitosan fragments and particulate impurities (Vakili et al., 2019). Then, sodium hydroxide solution is added to the filtrate, and the pH value is adjusted to be near the isoelectric point of chitosan (pH = 6.5) for flocculation and precipitation. The precipitate is dehydrated with ethanol and finally dried in an oven at 60°C to obtain white powder (Simon et al., 2025). This approach features maturity and stability, simple equipment requirements, and low production costs, and has become the mainstream in the current industry. However, it also shows obvious shortcomings. The production of one ton of chitosan produces 50 to 100 tons of strongly alkaline wastewater (pH > 12), which poses a major challenge to treatment and environmental pressure (Subramani et al., 2025). The product has wide molecular weight distribution (PDI > 3.0), high ash content (>1%), and poor batch uniformity. This seriously limits its application value in medicine and high-end biomaterials.

3.1.4. Defects of Traditional Preparation Ways

Traditional chitosan production methods have many drawbacks, mainly due to their heavy environmental burden. During the pretreatment stage, the processes of removing minerals and proteins respectively generate large amounts of acidic wastewater (pH < 2) and highly alkaline wastewater (pH > 12) (Berezin et al., 2025). The wastewater generated from such processes usually exhibits a chemical oxygen demand (COD) of over 20,000 mg/L, making it extremely difficult to biodegrade. The deacetylation process further exacerbates pollution, producing 50-100 tons of strongly alkaline wastewater for every ton of chitosan produced, with sodium hydroxide content of 4-5% and salt content exceeding 5% (Chattopadhyay et al., 2025). The wastewater needs to go through complex multi-stage neutralization, sedimentation and membrane separation treatments to meet the discharge standards, which greatly increases the environmental protection costs of enterprises. In addition, each ton of residue requires 3 to 5 tons of hydrochloric acid with a concentration of 30 to 35% and 8 to 12 tons of sodium hydroxide with a concentration of 50%. The consumption of acid and base reagents is terrifying. The raw material utilization rate is less than 60% (Alzahrani et al., 2025). Unreacted calcium salts and residual alkali solutions often cause the ash content of the product to exceed 1%, further affecting product quality.

Traditional methods show significant limitations in controlling molecular weight and ensuring product quality stability. Under high-temperature alkaline conditions, the β-(1,4)-glycosidic bonds in the chitosan main chain are extensively cleaved, causing the molecular weight to drop sharply from the raw material level of 100–1000 kDa to 5–50 kDa, with a corresponding loss rate exceeding 70% (Ashiq et al., 2025). Although sodium borohydride is usually added as a protective agent to reduce degradation by 15%, it cannot achieve precise control over molecular weight. This product presents a bimodal or multi-modal molecular weight distribution, with a polydispersity index (PDI) greater than 3.0 (Bernabé Vírseda et al., 2025). This leads to fluctuations of 20-30% in key properties such as viscosity and solubility between different batches. Moreover, high-temperature and strongly alkaline systems carry the risk of emitting volatile corrosive gases. Expensive alloys are required to be used for the equipment, which greatly increases the maintenance cost (Brahimi et al., 2025). The degree of deacetylation (DD) of the product is negatively correlated with the molecular weight (MW). Products with a high degree of deacetylation (>95%) have a low molecular weight (<50 kDa), which is difficult to meet the requirements of high molecular weight (>100 kDa) for applications such as tissue engineering scaffolds (Gadiock et al., 2025). Although the ash content and heavy metal residue levels comply with pharmacopeial standards, they still fail to meet the requirements for electronic-grade materials. This shortcoming severely restricts their application potential in biomedicine and high-end biomaterials.

3.2. Modern Preparation Methods

Traditional chitosan preparation techniques have long relied on extreme conditions such as strong acids, strong bases, high temperatures, and high pressures. This has resulted in severe molecular chain breaks, uncontrollable by-product pollution, and a limited range of product functions. Modern green technology, through interdisciplinary innovation, has established a precise control system for the entire process from raw material pretreatment to product functionalization. This system reduces environmental impact while enabling customized production of key parameters such as the polydispersity index (PDI) and deacetylation degree (DD) of chitosan molecules.

3.2.1. Biotechnology

The preparation of chitosan by enzymatic deacetylation driven by biotechnology has opened up a new path. The technology utilizes chitosan deacetylase (CDA) derived from fungi, such as Aspergillus oryzae, which operates under relatively mild conditions. The reaction can be completed within 24 to 72 h at pH 5.0 to 8.0 and a temperature of 40 to 50°C (Farid et al., 2025; Ianniciello et al., 2025). This process has obvious advantages. Notably, the resultant products exhibit a molecular weight retention rate exceeding 90%, a molecular weight range of 100–500 kDa, homogeneous deacetylation levels, and a polydispersity index (PDI) below 1.5. (Çatal et al., 2025). It completely avoids the problem of alkali pollution caused by the traditional alkali method, making the green preparation of chitosan possible.

However, enzymatic deacetylation also shows certain limitations. Currently, cost of enzymes is relatively high, accounting for about 30% of the production cost, greatly hindering their wide application. Moreover, the current technology can only achieve a maximum deacetylation degree of around 85%, which is insufficient for some high-demand applications (Benashvili et al., 2025). To overcome these challenges, researchers are working on enhancing enzyme activity through genetic engineering to increase the deacetylation limit and broaden its application scope. With the advancement of technology, enzymatic acetylation, with its advantages of being green, efficient and precise, is expected to become a key method for the industrial production of chitosan, promoting the industry towards environmental protection and sustainable development.

3.2.2. Microbial Pretreatment

Microbial pretreatment provides a green and environmental-friendly way for chitosan preparation. The principle is to use microbial metabolic activities to replace traditional chemical treatment steps. For example, during the fermentation of shrimp shells by lactic acid bacteria, the lactic acid produced can dissolve the calcium carbonate in the shrimp shells. Subsequently, the use of proteases to remove proteins can reduce the consumption of chemical reagents by 60% (Aljawish et al., 2025). Additionally, the biochemical oxygen demand/chemical oxygen demand (BOD_5/COD) ratio of the wastewater is greater than 0.4. It exhibits favorable biodegradability, facilitates subsequent treatment, and effectively mitigates the risk of environmental pollution. (Chattopadhyay et al., 2019; Cignolo et al., 2025).

In practical application, microbial pretreatment should consider microbial species, fermentation conditions and other factors. Different microorganisms have different metabolic abilities, so the appropriate strains should be selected according to the characteristics of raw materials. Factors like temperature, pH value, and aeration rate can also impact the effectiveness of pretreatment. The reasonable optimization of these parameters can enhance the efficiency and quality of the pretreatment process. With the enhancement of environmental awareness, the application prospects of microbial pretreatment technology are broad. In the future, more efficient and stable processes are expected to be developed, providing strong support for the green production of chitosan and promoting sustainable industrial development.

3.2.3. The Physical Field Strengthening Process

Microwave-assisted deacetylation is a major example of a physical field enhancement process. In a 15% to 30% sodium hydroxide solution, the microwave power is set at 300 to 800 W, the temperature range is 60 to 100°C, and the reaction time is only 10 to 40 min (Abinesh et al., 2025). Compared with traditional ways, this way significantly improves efficiency, shortens the reaction time by 80%, and reduces the amount of alkali used by half. It greatly reduces production costs and environmental pollution risks. The deacetylation degree of this product can reach 85% to 93%, and the molecular weight loss is less than 20% (da Silva Pinho-Jr et al., 2025). This way better preserves the molecular weight of the product compared to the 70% loss rate in traditional ways, ensuring high-quality chitosan production.

Currently, researchers are working to optimize the structure and performance of microwave reactors to enhance microwave energy efficiency and reaction uniformity. Studies are also underway to investigate the synergistic interactions between microwave irradiation and alternative technologies, with the goal of boosting preparation efficiency and refining product quality through these combined effects. With continuous efforts, the microwave-assisted deacetylation technology is expected to become the leading way for the green production of chitosan, promoting the sustainable development of the industry and providing a strong impetus for the large-scale production of chitosan.

3.2.4. Ultrasonic Assisted Extraction

Ultrasonic-assisted extraction plays a crucial role in the raw material pretreatment of chitosan preparation (Figure 4). The cavitation effect of ultrasound effectively disrupts the crystalline structure of chitosan in shrimp shells, reducing its crystallinity and thus accelerating the subsequent deacetylation reaction. Generally, ultrasonic pretreatment at 20-40 kHz can increase the mineralization rate by 40% and reduce the time required by 50%, significantly enhancing the efficiency of the pretreatment process (Cibotaru et al., 2025).

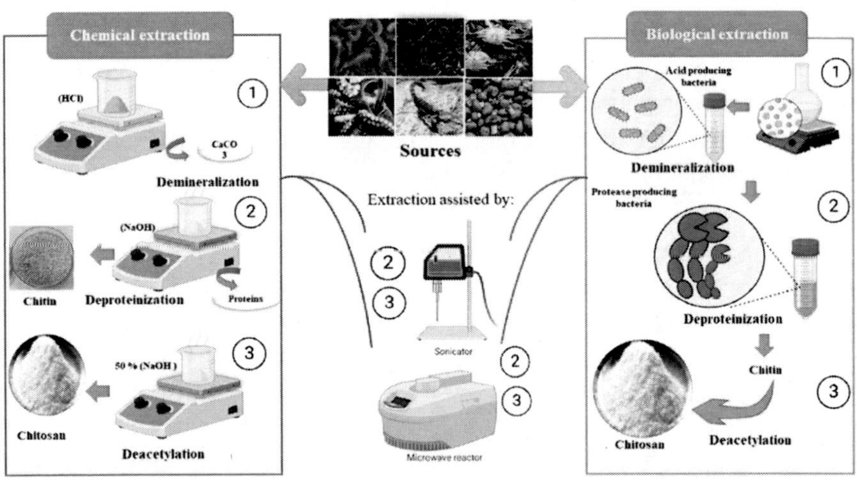

Figure 4. Ultrasonic and microwave-assisted chemical extraction and biological processes [Reproduced with permission from Elsevier, Ref. (Xu et al., 2025)].

The optimization of parameters in ultrasonic-assisted extraction technology is crucial for achieving its optimal performance. The destructive effects of ultrasonic waves at different frequencies on chitosan structures are various, and precise adjustments should be made based on the characteristics of the raw materials and production goals. Moreover, ultrasonic treatment can enhance the solubility and fluidity of chitosan solutions, facilitating the combination of chitosan with drug molecules and bioactive substances. With the widespread application of this technology, it is expected to play a more significant role in the green and efficient development of the chitosan industry, and drive the industry toward the direction of environmental protection, energy conservation, and quality enhancement.

3.2.5. Green Solvents and Process Innovation

Ionic liquids, as green solvents, have shown significant potential in the innovation of chitosan preparation. In the ionic liquid, 50% sodium hydroxide is added, and the reaction is carried out at 60°C for 2 h. The deacetylation degree of chitosan reaches over 90%, and the molecular weight retention rate increases by 50% (Jakubska et al., 2025). Additionally, the recovery rate of ionic liquids exceeds 95%, and wastewater is reduced by 90%, effectively alleviating environmental pressure and achieving a green and sustainable preparation (Álvarez-García et al., 2025).

Although ionic liquids are expensive, the understanding of their mechanisms is still incomplete. However, researchers are striving to reduce costs through structural design and synthesis ways and are using advanced analytical techniques to gain a deeper understanding of their mechanisms. In the future, the combination of green solvents and process innovation will trigger a green revolution in chitosan preparation, driving the industry towards more environmental-friendly and efficient practices. This will provide technical references for the green processing of biomass resources and guide the industry towards sustainable development.

3.2.6. The Mechanical Chemistry Method

The mechanical chemical method offers a solvent-free, green alternative for chitosan preparation. Through solid-state ball milling of chitosan and sodium hydroxide at 400 rpm for 2 h, this solvent-free reaction achieves a deacetylation degree of 75%, while maintaining a molecular weight above 100 kDa and a weight loss rate below 15% (Iñiguez-Moreno et al., 2025; Karpiński et al., 2025). This method avoids environmental and safety issues related to solvent use, reduces production costs, simplifies the process, and has broad application prospects in small-scale and decentralized preparation scenarios.

During the ball milling process, the impact and friction of the grinding balls refine chitosan particles, increasing their specific surface area. Mechanical forces also cause molecular chain distortions, enhancing reactivity and promoting deacetylation. However, the heat generated by ball milling can affect molecular structure, necessitating optimized temperature control in the process. In addition, the separation and purification of products are key factors (Jiang et al., 2019). With advancements in equipment and technology, mechanochemical ways are expected to achieve industrial production, complementing other technologies to develop composite preparation processes. This will further improve chitosan quality, promote the development of green preparation techniques, and meet the needs of various application.

3.2.7. Defects of Modern Preparation Methods

While modern preparation approaches have greatly promoted the green and efficient development of the chitosan industry, they are still confronted with multiple challenges. With regard to biotechnology, enzymatic deacetylation technologies show the advantages of mild reaction conditions and uniform product quality. However, the high cost of enzymes limits their large-scale industrial application. Additionally, the current upper limit of deacetylation is around 85%, which is insufficient for applications requiring higher deacetylation levels (Islam et al., 2025; Rehmat et al., 2019). Ionic liquids, as green solvents, show great potential in the preparation of chitosan. However, their high cost and incomplete understanding of their mechanism have limited wide application thereof. The mechanochemical way provides a solvent free and green option for the preparation of chitosan, but the heat during ball milling can affect the molecular structure. The separation and purification of products remain critical factors that usually lead to difficulties in purification. Moreover, microwave-assisted deacetylation and ultrasonic-assisted extraction have improved the preparation efficiency. But they require specialized equipment and generally lead to issues such as energy consumption and noise pollution (Karadirek et al., 2025). In short, modern preparation methods still face many challenges in terms of cost, technology, process, equipment and environmental impact, which require further research and optimization.

3.3. A Comparison Between Traditional Preparation and Modern Preparation Methods

3.3.1. Environmental Load

Traditional chitosan production methods impose a heavy environmental burden. The core steps include extraction of chitosan with strong acid and alkali and deacetylation reaction. These chemicals not only consume a lot, but also produce high concentrations of salts, organic matter and heavy metal wastewater after reaction. In addition, wastewater has extremely high chemical oxygen demand (COD) and biological oxygen demand (BOD), which makes it difficult to treat and easy to pollute water and soil for a long time (Knorr, 2025). In addition, the traditional high-temperature treatment step has further increased energy consumption and carbon emissions, aggravating the environmental load. However, modern preparation methods pay more attention to environmental protection and sustainability. Mainly through the use of biotechnology, green solvents and physical field enhancement technology, the dependence on chemical reagents is greatly reduced. These methods not only reduce the environmental burden but also meet the current demand for green and low-carbon production methods.

3.3.2. Product Quality

The traditional preparation methods often lead to the breaking and degradation of chitosan molecular chain due to the harsh reaction conditions. This issue induces wide molecular weight distribution, elevated polydispersity index values, and poor product quality stability. In contrast, modern preparation methods can accurately control the reaction conditions and accurately adjust key parameters such as molecular weight and deacetylation degree (Kritchenkov et al., 2019). For example, enzyme deacetylation can achieve precise control of deacetylation under mild conditions to prevent excessive chain breakage, so as to obtain high-quality chitosan products with narrow molecular weight distribution and low PDI (Koral et al., 2025). In addition, modern methods focus on the purification and post-treatment of products to further remove impurities and residual reagents, so as to improve the safety and reliability of products.

3.3.3. Production Cost

From the perspective of direct cost, the traditional preparation method has relatively low initial investment due to its mature process and simple equipment. However, considering the environmental cost, raw material utilization and product quality, the overall production cost is not low. The consumption of chemical reagents, the high cost of wastewater treatment and the instability of product quality all lead to the increase of subsequent treatment and application costs (kolamaki and mohajerani, 2025). All these lead to the high cost of traditional methods. While modern preparation approaches demand considerable upfront capital investment, they exhibit notable long-term cost benefits by improving raw material utilization efficiency, cutting down on chemical reagent usage and wastewater emissions, and yielding high-value-added products.

3.3.4. Technical Difficulties

The traditional preparation method has low technical threshold and is relatively easy to master and promote. With the increasingly stringent environmental regulations and the intensification of market competition, the technical limitations of the traditional process are becoming more and more obvious. These methods are difficult to meet the strict quality requirements of high-end applications, and the environmental pollution in the production process is also prominent (kulczy ń ska et al., 2025). In general, modern preparation methods integrate interdisciplinary technologies such as biotechnology, chemical engineering and materials science, making them more technically challenging.

3.3.5. Application Areas

Chitosan produced by traditional methods has unstable quality and low purity, which is mainly used in low-end markets such as agriculture and environmental protection. These industries have relatively low-quality requirements for chitosan, and the traditional process can meet their basic needs. At present, the demand for chitosan in high-end applications is also increasing. Modern chitosan production methods are widely applied in high-end fields including medicine, biomaterials, and food, owing to their superior product quality and performance. Especially in the pharmaceutical industry, chitosan prepared by modern preparation ways can be used to produce high-end products such as drug carriers, biomedical materials, and sustained-release agents. In biomaterials, chitosan can be used to produce tissue engineering scaffolds and materials for regenerating damaged tissues. These high-end applications expand the market and also enhance the industrial value.

Table 1. Comparison between traditional and modern preparation methods

Item	Traditional preparation	Modern preparation
Environmental load	Strong acid/alkali, high-temperature, energy-intensive, and pollution	Biological enzymes, green solvents, less chemical reagents, less wastewater, more environmental-friendly.
Product quality	Broad molecular weight, distribution, instability, and low purity	Precise control of reaction, narrow molecular weight distribution, precise deacetylation degree, stability and high purity.
Production cost	Low initial investment, yet high reagent use, wastewater costs, and overall expenses	High initial investment, efficient resource use, low long-term costs
Technical difficulty	Low technical threshold, yet high-end demand challenges and prominent pollution	Diffusion lowers barriers; Multi-disciplinary tech integration challenges
Application	Low-end market application	High-end market application

4. Quality Control of Chitosan

4.1. Raw Material Control: The Whole Chain Traceability System

The quality control of chitosan requires the establishment of a complete traceability system from raw material collection to finished products. The quality of chitosan as a raw material directly determines the performance limit of the final product, so strict screening standards should be implemented. High quality shrimp and crab shells should meet the following key indicators (Wagener et al., 2025). The chitosan content is greater than 20% to ensure extraction efficiency and cost-effectiveness. The ash content is strictly controlled below 2% to reduce the burden of subsequent purification processes. The residual amount of heavy metals should comply with national standards to ensure the biological safety of the product, such as lead content not exceeding 1.0 mg/kg and arsenic content not exceeding 0.3 mg/kg (Wang et al., 2025c; Yang et al., 2019). In the pretreatment stage, "three-step" is the key. Firstly, protein impurities can be effectively removed by treating with 3% (w/v) sodium hydroxide solution at 80°C for 2 h. Secondly, it was treated with 10% (v/v) hydrochloric acid solution for 12 hours at room temperature to remove calcium and dissolve inorganic salts such as calcium carbonate (Wang et al., 2025a; Zeng et al., 2025). Ultrasonic assisted treatment can significantly improve the purity of chitosan and shorten the treatment time. The storage stage is also crucial, which requires strict control of warehouse humidity within the range of 40–60% relative humidity (RH) and temperature between

15–25°C. (YıLDıRım et al., 2025). The use of double-layer polyethylene bags filled with high-purity nitrogen can effectively extend the shelf life of raw materials to 36 months and ensure stable quality.

4.2. Process Monitoring: Intelligent Response Control System

The preparation of chitosan needs to be accurately monitored by intelligent reaction control system. Deacetylation reaction is a key step in this process, and its conditions directly affect the degree of deacetylation and molecular weight distribution of the product. Optimization of critical parameters such as alkali concentration, reaction temperature, and reaction duration enables the deacetylation degree to reach 97% or higher, thus meeting the application criteria for intermittent alkali treatment in high-end medical biomaterials (Yakar et al., 2025). In addition, the near infrared spectroscopy technology can detect the degree of deacetylation quickly and nondestructively, shorten the detection time to 1 min, and the accuracy is ±0.5% (Wang et al., 2025b). Raman spectroscopy can monitor the reaction process in real time, and provide real-time data support for process adjustment through the change of characteristic peak intensity. In addition, by maintaining a constant temperature of 60-65°C and combining dynamic viscosity monitoring, the key parameters for controlling molecular weight and acid leaching time were optimized (Li et al., 2019b). High viscosity chitosan products with molecular weight over 1000 MPa·s can be obtained, which can meet the needs of high-end applications such as tissue engineering scaffolds.

4.3. Product Characterization: A Multi-Dimensional Detection System

A comprehensive characterization and evaluation of chitosan ought to be performed via a multidimensional testing platform. Physical and chemical properties include deacetylation degree, molecular weight distribution, ash content and other key indicators. The degree of deacetylation was determined by ^{1} H-NMR and calculated from the integral area ratio of characteristic peaks (ren et al., 2025). The molecular weight distribution was analyzed by gel permeation chromatography. The polydispersant index (PDI) value is required to be less than 1.5 to ensure the stability and consistency of product performance (Zhang et al., 2025). The ash content was determined by thermogravimetry. The ash content was less than 2% to reflect the purity of chitosan. Microbial limits are also important. The total number of aerobic bacteria is not more than 1000 cfu/g, and the total number of mold and yeast is not more than 100 cfu/g, which ensures the health, safety and biocompatibility of the product (Zhou et al., 2025). Analysis of functional properties is indispensable, which involves determining intrinsic viscosity via viscometry to assess the rheological performance of the resultant products. The characteristic peaks were verified by FTIR, and the chemical structure of chitosan was characterized. The crystallinity of the product was analyzed by X-ray diffraction (XRD) to understand the physical state of the product.

4.4. Standard System: International Mutual Recognition Quality Framework

The quality control of chitosan should follow the internationally recognized quality framework to ensure the global competitiveness of products. Domestic and international standards complement each other and provide double guarantee for product quality. In terms of domestic standards, gb/t 38479-2021 specifies the determination method of chitosan content to ensure that the product quality is controllable (Zhao et al., 2025). Internationally, the United States Pharmacopeia (USP) has strict requirements for heavy metal content, which does not exceed 10 ppm to ensure the purity of the product (Zhang et al., 2019). European Pharmacopoeia (EP) has set endotoxin limits consistent with those of the FDA regulations, requiring protein residues to be less than 0.3%. This further reduces the potential immunogenic risk. Enterprises are required to establish a three-level quality documentation system, covering raw material traceability records, process parameter logs, and inspection reports, to ensure compliance with the regulatory requirements of target markets. (Zhao and Saldaña, 2019). Through comprehensive quality control of the entire production process, the yield of high-quality chitosan products can be increased from 75% to over 92%. Enterprises create greater economic benefits and social value through strict quality control, and at the same time promote the development of the chitosan industry towards high-end and green directions.

Conclusion and Challenges

Chitosan is a natural polysaccharide endowed with excellent biocompatibility, biodegradability, and metal ion complexation capacity, which exhibits significant application potential in biomedicine, environmental protection, and modern agriculture. Current studies are driving the chitosan industry towards a more efficient, environmentally friendly and intelligent direction through the development of multi-source raw materials, including crustacean waste, fungal cell walls, insect exoskeletons and synthetic biology pathways. The advancement is supported by innovative green preparation technologies, such as enzyme deacetylation, physical field enhancement, and ionic liquid systems. It is also supplemented by a comprehensive quality control framework, including full-chain traceability, intelligent process monitoring and multi-parameter characterization. Key breakthroughs include the realization of customized synthesis in microbial cell factories, an 80% reduction in alkali consumption through microwave, and an AI-driven reaction model achieving a ±2% deacetylation control accuracy. These developments collectively position chitosan as a key material in the bioeconomy era, providing sustainable solutions through technological innovation.

The future of chitosan industry largely depends on the integration of multidisciplinary technology. Gene editing uses crispr-cas9 system to reshape the fungal fermentation system, accurately control the expression of chitosan synthase and deacetylase in *Aspergillus strains*, and realize the synchronous optimization of deacetylation degree and molecular weight. At the same time, the continuous flow reactor driven by artificial intelligence overcomes the limitations of mass production. Machine learning algorithms were employed for real-time optimization of parameters including microwave power and ultrasonic frequency, enabling the deacetylation degree to be precisely controlled within a margin of ±1%. The intelligent manufacturing mode can increase the utilization rate of raw materials to more than 95% and significantly reduce the formation of by-products. More notably, the combination of chitosan and nanotechnology has paved the way for promising new application avenues. For example, compared with traditional dressings, smart hydrogels loaded with growth factors can achieve pH responsive release and show superior repair efficiency in the treatment of diabetic ulcer. In addition, chitosan has great potential in sustainable development and resource recycling. Chitosan flocculants are transitioning from laboratory to industrial application. They have the advantages of high removal rate of heavy metals and low sludge yield in wastewater treatment.

Despite the promising prospects of the chitosan industry, it still faces multiple technical bottlenecks and industrialization challenges. Regarding process optimization, how to achieve accurate regulation of deacetylation degree (DDA) and molecular weight (MW) continues to be a key challenge. The traditional concentrated alkali way can achieve a relatively high DDA (>95%), but its loss rate of MW exceeds 70%. The enzymatic deacetylation method is limited by the high cost of CDA enzymes (over 30% of production costs) and the upper limit of DDA (about 85%) while maintaining molecular weight. In large-scale production, microbial fermentation processes are limited by the stability of strains and fermentation efficiency. This leads to the chitosan yield in industrial bioreactors being lower than 5g/L, significantly lower than that of chemical ways. Moreover, from the perspective of environmental sustainability, the traditional production of 1 ton of chitosan generates 50–100 tons of high-alkalinity wastewater with high chemical oxygen demand (COD), and the associated treatment costs account for over 25% of the total production cost. Therefore, it is necessary to develop zero-discharge technology. Market access and standardization are another major obstacle. The 32 major economies around the world have implemented differentiated quality supervision systems for chitosan. For instance, the EU Pharmacopoeia requires bacterial endotoxin to be less than 20 EU/mg, and the FDA standard limits protein residue to less than 0.3%. This leads to enterprises having to invest an average of 8 million US dollars annually in cross-border certification. More critically, although ISO/TC348 has initiated global standard-setting for chitosan, there are still disagreements on core technical indicators such as DDA measurement ways and MW distribution characterization, increasing the risk of technical trade barriers

Acknowledgments

This chapter was supported by Talent Recruitment Program of Shaanxi University Technology [grant number: SLGRCQD037, X20250004], and the Shaanxi Provincial Science and Technology Department Project [grant number: 2024JC-YBQN-0835].

References

Aberoumand, A., Hoseinian, M. (2025). Extraction of Chitosan from shells of crab (Liocarcinus vernalis). *Applied Food Research*, 5, 100964.

Abinesh, A. R., Yuvaraaj, S. A., Subashini, S., Pravin, R., Baskar, G. (2025). Phytosynthesis of ZnO-Chitosan nanobiocomposite biofunctionalized using fungal L-Asparaginase for the treatment of human lung cancer. *Journal of Molecular Structure*, 143174.

Aljawish, A., Habay, N., Roby, M. H. H. (2025). Eco-friendly synthesis of chitosan derivatives for sustainable packaging materials: Enhanced antioxidant and barrier properties. *Materials Today Communications*, 46, 112782.

Álvarez-García, S., D'Isita, I., Pistillo, O. M., Moumni, M., Germinara, G. S., Romanazzi, G. (2025). Volatile organic compounds released by chitosan formulations present diverse chemical composition and produce differential effects on postharvest pathogens. *Postharvest Biology and Technology*, 228, 113619.

Alzahrani, S. O., Alisaac, A., Alsahag, M., Aldosari, F. M., Alshammari, N. M., Alhalafi, M. H., Obaid, A. O., El-Metwaly, N. M. (2025). Green synthesis, characterization, phytochemical analyses, and antibacterial evaluation of Ag/TiO2 and Ag/TiO2-chitosan nanocomposites derived from M. chamomilla. *International Journal of Biological Macromolecules*, 312, 144220.

Ashiq, A., Das, S. C., Biswas, S., Akter, R., Masum, M. A., Bedoura, S. (2025). Accelerating wound healing: Unveiling a synergistic neem-chitosan composite for rapid tissue regeneration. *Next Research*, 2, 100541.

Basso, L., Vuillet, C., Rahbé, Y., David, L., Charlot, A., Sudre, G. (2025). Chitin-binding protein behavior at chitosan interface studied by quartz crystal microbalance with dissipation monitoring (QCM-D): Binding quantification, orientation and affinity constants. *Colloids and Surfaces B: Biointerfaces*, 252, 114650.

Benashvili, M., Lee, S., Ju, Y.-W., Park, Y.-J., Kim, S.-A., Park, S.-W., Son, Y.-J. (2025). Effect of mealworm (Tenebrio molitor L.) chitosan coating on the postharvest qualities of strawberries. *Postharvest Biology and Technology*, 228, 113657.

Berezin, B. B., Kulikov, S. N., Zhuravleva, I. L., Ezernitskaya, M. G., Piskarev, V. E., Antonov, Y. A., Bezrodnykh, E. A., Tikhonov, V. E. (2025). Complexation with chicken yolk phosvitin modulates antibacterial activity of chitosan. *Carbohydrate Polymers*, 362, 123691.

Bernabé Vírseda, I., de la Orden, M. U., Martínez Urreaga, J. (2025). Effects of chitosan thin barrier layers on the oxygen permeability and optical properties of poly(lactic acid) and poly(3-hydroxybutyrate-co-3-hydroxyvalerate) multilayers. *Carbohydrate Polymer Technologies and Applications*, 11, 100918.

Brahimi, I., Allali, K., Benadjila, A., Zamoum, M., Goudjal, Y. (2025). Development and optimization of a biological coating process based on chitosan from edible cricket to control Aspergillus niger-decay and weight loss of the Algerian dates cv. Deglet Nour. *Process Biochemistry*, 157, 21-32.

Çatal, T., Bayrak, A., Çanlıoğlu, A., Çakmur, B. B., Tıngır, M., Kural, A., Gülçubuk, A., Öztürk, G. Y., Kural, C., Duramaz, A. (2025). The role of chitosan in tissue healing after primary repair of quadriceps tendon ruptures: Experimental animal model. *The Knee*, 55, 193-205.

Chattopadhyay, K., Xavier, K. A. M., Layana, P., Balange, A. K., Nayak, B. B. (2019). Chitosan hydrogel inclusion in fish mince-based emulsion sausages: Effect of gel interaction on functional and physicochemical qualities. *International Journal of Biological Macromolecules*, 134, 1063-1069.

Chattopadhyay, K., Xavier, K. A. M., Porayil, L., Balange, A. K., Nayak, B. B. (2025). Effects of chitosan molecular weight on proteins and lipids interactions in fish mince emulsion sausages. *International Journal of Biological Macromolecules*, 319, 145562.

Chevrier, A., Wang, D., Ladd, E., Blais, J.-S., Lavertu, M. (2025). Chitosan powders and Ortho-R, a chitosan-containing finished product, are stable long term when stored in proper conditions. *Carbohydrate Polymer Technologies and Applications*, 10, 100777.

Cibotaru, S., Anisiei, A., Platon, V.-M., Rosca, I., Sandu, I.-A. Coman, C.-G., Mititelu-Tartau, L., Andreica, B.-I., Marin, L., (2025). Imino-quaternized chitosan/chitosan nanofibers loaded with norfloxacin as potential bandages for wound healing. *International Journal of Biological Macromolecules*, 314, 144304.

Cignolo, D., Rizzi, V., Fini, P., Petrella, A., Milanese, C., Cosma, P., Gubitosa, J. (2025). Composite Chitosan-Biochar-based sponges for the removal of Direct Blue-78 from water: Physical and chemical parameters affecting the adsorption process. *Colloids and Surfaces A: Physicochemical and Engineering Aspects*, 724, 137518.

da Silva Pinho-Jr, J., Moreira, B. S., Brito, G. B., de Almeida Torres, L. G., Eiriz, D. N., Torres, A. G., Teodoro, A. J., Perrone, D., Castelo-Branco, V. N. (2025). Impact of chitosan-based oleogels crosslinked with vanillin containing different oil phases on lipid digestion and bioaccessibility of carotenoids and tocols. *Food Chemistry*, 487, 144740.

da Silva Ponciano, C., Gonzaga, F. C., de Oliveira, C. P. (2025). Smart packaging based on chitosan acting as indicators of changes in food: A technological prospecting review. *International Journal of Biological Macromolecules*, 145722.

Dejene, B. K., Abtew, M. A., (2025). Chitosan/zinc oxide (ZnO) nanocomposites: A critical review of emerging multifunctional applications in food preservation and biomedical systems. *International Journal of Biological Macromolecules*, 316, 144773.

Deshmukh, S. S., Nijhawan, H. P., Yadav, K. S. (2025). Chitosan-based copolymeric nanocarrier systems: A novel paradigm for targeted breast cancer therapy. *International Journal of Biological Macromolecules*, 319, 145537.

Dong, Z., Jiang, X., Jiang, Z., Lv, L., He, M. (2019). Preparation of glycerol plasticized chitosan films using AlCl3·6H2O as the solvent: optical, crystalline, mechanical and barrier properties. *International Journal of Polymer Analysis and Characterization*, 24, 295-303.

Drebezghova, V., Cugnet, C., Fernandes, S. C. M., Dupin, J.-C., Guignard, M., Ranchou-Peyruse, A., Ranchou-Peyruse, M., Nardin, C. (2025). Chitosan-based coatings to prevent Nitratidesulfovibrio vulgaris influenced (bio)corrosion on aluminium alloy. *Colloids and Surfaces B: Biointerfaces*, 253, 114751.

El-Naggar, N. E. A., Elshafey, N., Alafifi, H. I., Eltahy, M. A., Haikl, R. I., ElShazly, H. A., Ahmed, Y. W., Hassan, H. I., Safo, M. M., Haroun, S. A., El-Sawah, A. A. (2025). Innovative strategy for chitosan nanoparticles biosynthesis using Gelidium amansii, statistical optimization, characterization, cytotoxicity and molecular docking against hepatocellular carcinoma. *International Journal of Biological Macromolecules*, 311, 143687.

El-Saadony, M. T., Saad, A. M., Alkafaas, S. S., Dladla, M., Ghosh, S., Elkafas, S. S., Hafez, W., Ezzat, S. M., Khedr, S. A., Hussien, A. M., Fahmy, M. A., Elesawi, I. E., Salem, H. M., Mohammed, D. M., Abd El-Mageed, T. A., Ahmed, A. E., Mosa, W. F. A., El-Tarabily, M. K., AbuQamar, S. F., El-Tarabily, K. A. (2025). Chitosan, derivatives, and its nanoparticles: Preparation, physicochemical properties, biological activities, and biomedical applications–A comprehensive review. *International Journal of Biological Macromolecules*, 313, 142832.

Farid, A., Hossam, A., Waheed, M., Safwat, G. (2025). Synthesis of rice husk-loaded chitosan nanoparticles for treatment of amikacin-induced kidney injury in male rats. *International Journal of Biological Macromolecules*, 312, 144125.

Franconetti, A., Carnerero, J. M., Prado-Gotor, R., Cabrera-Escribano, F., Jaime, C. (2019). Chitosan as a capping agent: Insights on the stabilization of gold nanoparticles. *Carbohydrate Polymers*, 207, 806-814.

Gadiock, K. S., Manogaran, Y., Ramasamy, P. (2025). Preparation of chitosan sulfate from marine cuttlefish Sepia prashadi (Winckworth, 1936) cuttlebone and its anti-blood clotting efficacy. *Next Research*, 2, 100530.

Ghorai, P., Jana, B., Ghorai, S., Ganguly, J. (2025). Synthetic strategy for multimodal –NH2 functionalized chitosan-based materials towards sustainable evolution. *Carbohydrate Research*, 552, 109482.

Gou, D., Ren, P., Ren, F., Wang, Q., Zhang, H., Cheng, X., Wang, L., Liu, T., Zhao, J. (2025). Fabrication, characterization and food applications of multifunctional chitosan-based aerogels: A review. *International Journal of Biological Macromolecules*, 145702.

Han, X., Tian, L., Chen, Q., Wang, L., Mi, Y., Li, Q., Guo, Z., Dong, F. (2025). Preparation of novel thiolated chitosan with significant antioxidant activity. *International Journal of Biological Macromolecules*, 308, 142536.

He, J., Xi, J., Han, B., Xiao, Q., Guo, D., Huang, S. (2025). Chiral stationary phase based on chitosan Schiff base derivatives for enantioseparation. Carbohydrate Polymers 359, 123590.

Hill, C. T., Tabor, R. F. (2025). Properties of scalable, pH-sensitive chitosan polymer-stabilised emulsions. *Journal of Molecular Liquids*, 433, 127888.

Huang, Y., Dabbour, M., Mintah, B. K., Pan, J., Wu, M., Long, R., Zhang, S., Liu, S., Liao, H., He, R., Ma, H. (2025). Extraction and characterization of chitin and chitosan from molts and Eupolyphaga sinensis Walker: Influence of chemical and ultrasound-assisted enzymatic processing. *International Journal of Biological Macromolecules*, 309, 142804.

Ianniciello, D., Montosa, A. P., de Melo Barbosa, R., Villén, F. G., Salvia, R., Scieuzo, C., Viseras, C., Falabella, P. (2025). Development of chitosan-clay nanocomposite films from Hermetia illucens: analysis of chemical, physical, and mechanical properties. *International Journal of Biological Macromolecules*, 311, 143496.

Ibrahim, A. G., Elgammal, W. E., El-Sayed Ebead, E., Ismail, M. F., Sharaf, M. H., Aboelnaga, A. (2025). Functionalized chitosan with new acrylohydrazide derivative: Synthesis, characterization, DFT calculations, antimicrobial and antiproliferative activities. *Journal of Molecular Structure*, 1341, 142607.

Iñiguez-Moreno, M., Hernández-Varela, J. D., Burelo, M., Elizondo-Luevano, J. H., Araújo, R. G., Treviño-Quintanilla, C. D., Medina, D. I. (2025). Progress in chitosan-based materials: Enhancing edible coatings and films through modifications and functionalization for food preservation. *Process Biochemistry*, 156, 175-190.

Islam, D., Aye, T. T., Islam, M. T., Saha, D., Majumdar, P. R., Bhowmik, S., Mamun, A.-A. (2025). Chitosan gel in extending the shelf-life of Nga-pi: A fermented fish product of Bangladesh. *Food and Humanity*, 5, 100663.

Jakubska, J., Wawoczny, A., Kluska, D., Grzybek, P., Hudecki, A., Kołodziej, M., Jasiak-Jaroń, K., Abdullah, Shyntum, D. Y., Gołombek, K., Pakieła, W., Krukiewicz, K., Gillner, D., Dudek, G. (2025). Dual functionality of deep eutectic solvents in chitosan-based biodegradable films for food packaging applications. *Industrial Crops and Products*, 232, 121268.

Jiang, Z., Song, Y., Qiao, J., Yang, Y., Zhang, W., Liu, W., Han, B. (2019). Rat sciatic nerve regeneration across a 10-mm defect bridged by a chitin/CM-chitosan artificial nerve graft. *International Journal of Biological Macromolecules*, 129, 997-1005.

Karadirek, S., Turan, B., Demircivi, P. (2025). Construction of $BaZrO_3$/chitosan composite for sufficient antibiotic adsorptive removal: Central composite design analysis and antimicrobial performance. *Journal of the Indian Chemical Society*, 102, 101813.

Karpiński, R., Łukaszczyk, A., Marchewka, J., Błoniarz, A., Kąc, S., Kopia, A., Sitarz, M., Drożdż, K., Biegun-Drożdż, K., Gosiewski, T., Brzychczy-Włoch, M., Moskalewicz, T. (2025). Effect of cellulose type on microstructure and properties of cellulose/curcumin/chitosan coatings obtained on titanium substrates by electrophoretic deposition. *Applied Surface Science*, 709, 163788.

Khani, F. M. R., Naseri, M., Babaei, S., Hosseini, S. M. H., Valizadeh, S. (2025). Cumin essential oil-infused chitosan-PVA composite coatings: An innovative approach for developing antimicrobial, antioxidant and functionalized polyethylene films. *International Journal of Biological Macromolecules*, 145270.

Knorr, D., (2025). Food processing of chitin and chitosan: From waste to opportunities. *Journal of Food Engineering*, 400, 112634.

Kolamaki, A. M., Mohadjerani, M. (2025). Preparation, characterization and in vitro bioactivity evaluation of quercetin and syringic acid loaded colloidal nanoparticles of chitosan-caseinate complex. *Carbohydrate Polymer Technologies and Applications*, 11, 100904.

Koral, C., Marsico, M., Mazaheri, Z., Papari, G. P., Andreone, A., Teghil, R., De Bonis, A. (2025). Enhanced relaxation dynamics in silver-doped chitosan nanocomposites: Probing the charge carrier dynamics and terahertz dielectric response. *Materials Chemistry and Physics*, 344, 131137.

Kritchenkov, A. S., Egorov, A. R., Volkova, O. V., Kritchenkov, I. S., Kurliuk, A. V., Shakola, T. V., Khrustalev, V. N. (2019). Ultrasound-assisted catalyst-free phenol-yne reaction for the synthesis of new water-soluble chitosan derivatives and their nanoparticles with enhanced antibacterial properties. *International Journal of Biological Macromolecules*, 139, 103-113.

Kulczyńska, J., Chibrikov, V., Pawcenis, D., Jędrzejczyk, R. J., Sebastian, V., Cybulska, J., Zdunek, A., Kyzioł, A. (2025). Sustainable poly(vinyl alcohol)/chitosan electrospun nanofibers and casted films with bioactive additives – comparative study on physicochemical properties and in vitro biological activity. *Industrial Crops and Products*, 233, 121335.

Kumar, D., Khullar, P., Jain, V., Prakash, R., Rai, B. (2025). Investigating the corrosion inhibition mechanism of chitosan for steels in acidic environments: Experimental and computational studies. *Journal of Molecular Structure*, 1337, 142057.

Lambrianidi, L., Savvaidis, I. N., Tsiraki, M. I., El-Obeid, T. (2019). Chitosan and oregano oil treatments, individually or in combination, used to increase the shelf life of vacuum-packaged, refrigerated european eel (*Anguilla anguilla*) fillets. Journal of Food Protection, 82, 1369-1376.

Lemine, O. M., Elansary, M., Algessair, S., Madkhali, N., Ali, R., Oulhakem, O., Mouhib, Y., Alshammari, M., Alsubaie, S. M., Belaiche, M., Henini, M. (2025). Enhanced in-vitro magnetic hyperthermia performance of chitosan-coated $CoFe_2O_4$ nanoparticles. *Materials Today Communications*, 46, 112852.

Li, J., Liu, Y., Zhao, X. (2025). Latest research progress on antibacterial properties of chitosan-based nanofibers. *Chemical Engineering Journal*, 517, 163776.

Li, Q., Mao, Q., Yang, C., Zhang, S., He, G., Zhang, X., Zhang, W. (2019a). Hydrophobic-modified montmorillonite coating onto crosslinked chitosan as the core-shell micro-sorbent for iodide adsorptive removal via Pickering emulsion polymerization. *International Journal of Biological Macromolecules*, 141, 987-996.

Li, Q., Wei, L., Zhang, J., Gu, G., Guo, Z. (2019b). Significantly enhanced antioxidant activity of chitosan through chemical modification with coumarins11Electronic supplementary information (ESI) available. *Polymer Chemistry*, 10, 1480-1488.

Lieu, M. D., Nguyen, H. H., Huynh, N. A. T., Tuyen Ha, P. K., Dang, T. K. T., Nguyen, T.H. (2025). Green synthesized silver nanoparticles by avocado leaves incorporating chitosan edible coating for avocado preservation under ambient conditions. *Biocatalysis and Agricultural Biotechnology*, 67, 103650.

Lin, Y.-C., Ramanathan, S., Wang, H.-Y., Lin, Y.-C., Liu, W.-C., Jones, J. R., Cho, N.-J., Hu, C.-C., Chung, R.-J. (2025). Engineered bioactive glass-chitosan hybrid for dual tissue and bone regeneration multifunctional healing. *Biomaterials Advances*, 176, 214340.

Liu, C., Crini, G., Lichtfouse, E., Wilson, L. D., Picos-Corrales, L. A., Balasubramanian, P., Li, F. (2025). Chitosan-based materials for emerging contaminants removal: Bibliometric analysis, research progress, and directions. *Journal of Water Process Engineering*, 71, 107327.

Liu, J., Zhang, X., Kennedy, J. F., Jiang, M., Cai, Q., Wu, X. (2019). Chitosan induces resistance to tuber rot in stored potato caused by Alternaria tenuissima. *International Journal of Biological Macromolecules*, 140, 851-857.

Long, Y., Liu, X., Huang, F., Peng, L., Duan, Y., Bai, G., Zhu, Z., Su, D. (2025a). Normal-pressure-prepared chitosan carbonate liquid dressing: Spontaneous transformation into a pure chitosan water-resistant film and enhanced wound repair. *Carbohydrate Polymers*, 366, 123805.

Long, Y., Xu, W., Huang, F., Peng, L., Duan, Y., Su, D., Zhu, Z. (2025b). Spontaneous chitosan hydrogel formation through carbonic acid modulation: A crosslinker-free, efficient preparation for dressings. International Journal of Biological Macromolecules 319, 145618.

Lu, J., Yin, X., Wu, Q., Wang, M., Han, L., Su, S. (2025). Fabrication and stability of Pickering nanoemulsion stabilized by self-aggregated chitosan nanoparticles. *Food Chemistry*, 484, 144445.

Mandal, S., Hazra, C., Joshi, N., Kumar, P., Sen, R., Das, S., Das, K. (2025). Development and performance evaluation of chitosan hydrogel based ZnO/CuO antibacterial and antialgal nanocoating for the control of biofouling in cultural heritage stones. *Construction and Building Materials*, 489, 142236.

Manohar, M., Shetty, A., Lobo, C. L., Jain, P., Hebbar, S., Dhas, N., Sutar, K. P., Sukeewandhi, J., Perumalsamy, H., Balusamy, S. R., Jamous, Y. F., Dubey, A., El-Zahaby, S. A. (2025). Emerging trends in chitosan based colloidal drug delivery systems: A translational journey from research to practice. *Carbohydrate Polymers*, 360, 123604.

Martingo, M., Baptista-Silva, S., Mesquita, R., Ferreira, J. P., Borges, S., Pintado, M. (2025). Exploring the potential of mealworm chitosan for hemodialysis applications. *Sustainable Chemistry and Pharmacy*, 45, 102013.

Mawale, K. S., Giridhar, P., Johnson, T. S. (2025). Chitosan: A versatile polymer for enhancing plant bioactive accumulation, managing plant diseases, and advancing food preservation technologies. *International Journal of Biological Macromolecules*, 308, 142081.

Milbreta, U., Andze, L., Zoldners, J., Irbe, I., Skute, M., Filipova, I. (2025). Exploring the potential of locally sourced fungal chitosan for paper mechanical property enhancement. *Journal of Renewable Materials*, 13, 583-597.

Mohseni, M., Mousavi, M., Kiani, H., Homayoonfal, M. (2025). Chitosan-coated nanoliposomes as vehicles for l-citrulline: Preparation, characterization, antioxidant properties, and nanovesicles behavior. *Carbohydrate Polymer Technologies and Applications*, 11, 100925.

Moussa, A., Crépet, A., Ladavière, C., Trombotto, S. (2019). Reducing-end "clickable" functionalizations of chitosan oligomers for the synthesis of chitosan-based diblock copolymers. *Carbohydrate Polymers*, 219, 387-394.

Nabil, Y., Atta, A. H., El-Fattah, W. A., Hafez, H. S., Elshaarawy, R. F. M. (2025). Progress in chitosan/essential oil/ZnO nanobiocomposites fabrication for wound healing applications: A review. *International Journal of Biological Macromolecules*, 318, 145123.

Naveedunissa, S., Meenalotchani, R., Manisha, M., Ankul Singh, S., Nirenjen, S., Anitha, K., Harikrishnan, N., Prajapati, B. G. (2025). Advances in chitosan based nanocarriers for targetted wound healing therapies: A review. *Carbohydrate Polymer Technologies and Applications*, 11, 100891.

Oliveira Júnior, F. D., Pinheiro, A. C., Vicente, A. A., Cunha, R. L. (2025). Development of chitosan/navy bean extract-based nanocarriers for oral delivery of insulin. *Food Research* International, 218, 116876.

Olunusi, S. O., Ramli, N. H., Adam, F., Hossain, E., Kim, T. H. (2025). Chitosan as a sustainable alternative for fresh food packaging: Structural insights, modification strategies, and innovations for commercial viability. *International Journal of Biological Macromolecules*, 318, 145065.

Panda, T. K., Bollepalli, D., Beji, D. S., Varanasi, K., Ananth, M., Rajarajeswaran, J., Eswar Neerugatti, K. R. (2025). Beyond Slurry Systems: Chitosan Thin Film Fabrication, Mechanism, and Reusability for Sustainable Textile Effluent Remediation. *Separation and Purification Technology*, 368, 132936.

Patel, P. K., Uppaluri, R. V. S. (2025). Environmental sustainability through adsorption: A review of chitosan's potential in dye pollution remediation. *Sustainable Chemistry and Pharmacy*, 46, 102096.

Peng, Y., Yu, Y., Su, Z., Zhong, Y., Chen, Y., Mao, Y., Vijayakumar, S., Xin, M., Li, M. (2025). Synthesis and performance optimization of dual-modified chitosan Derivatives: Effects of novel zwitterionic and quaternary ammonium groups on antibacterial, antioxidant, and biocompatibility properties. *Carbohydrate Research*, 554, 109556.

Piboonkunsamlit, S., Suntara, C., Cherdthong, A. (2025). The latest insights and perspectives on chitosan supplementation driving ruminant feed efficiency and sustainability. *Animal Feed Science and Technology*, 323, 116298.

Popyrina, T. N., Aksenova, N. A., Shilkova, A. A., Minaeva, E. D., Agapov, I. G., Pal, K., Minaev, N. V., Timashev, P. S., Akopova, T. A., Demina, T. S. (2025). Emulsifying and stabilizing ability of alkylated chitosan. *International Journal of Biological Macromolecules*, 318, 145022.

Putra, P. S. U., Adhika, D. R., Asri, L. A. T. W., Suprijadi, S. (2025). Performance Enhancement of Chitosan for Food Packaging: Impact of Additives and Nanotechnology. *Journal of Renewable Materials*, 13, 1043-1070.

Putro, J. N., Bintang Ilhami, F., Cheng, C.-C., Imanuela, K. J., Mangindaan, D., Irawaty, W., Rahmawati, A., Shimomura, O., Ismadji, S. (2025). Polyelectrolyte nanoparticles derived from durian seed gum modified with chitosan for selective targeting in cancer cells. *International Journal of Biological Macromolecules*, 315, 144625.

Rajeswari, M. V., Suresh, A., Kumar, E. T., Subramanian, K., Kaliaperumal, K., Sampath, S., Devanesan, S., AlSalhi, M. S. (2025). Efficacy of chitosan conjugated bivalent vaccine on immune response and disease resistance against Aeromonas hydrophila, and Aeromonas caviae in Goldfish Carassius auratus. *Journal of Drug Delivery Science and Technology*, 107190.

Rani, K., Malik, A. K., Setia, A., Randhave, N. V., Verma, N., Kumar, V., Vaishali, Deshmukh, K., Muthu, M. S. (2025). Chitosan and its derivatives as nanotheranostics in multiple diseases management: a clinical perspective. *Carbohydrate Polymers*, 366, 123852.

Rehmat, Z., Mohammed, W. S., Sadiq, M. B., Somarapalli, M., Kumar Anal, A. (2019). Ochratoxin A detection in coffee by competitive inhibition assay using chitosan-based surface plasmon resonance compact system. *Colloids and Surfaces B: Biointerfaces*, 174, 569-574.

Ren, H., Xu, J., Lai, Y., Xu, R., Li, J., Shen, J.-W., Chen, J.-X. (2025). pH sensitive loading and release of doxorubicin by chitosan-graphene quantum dots hybridized material. *International Journal of Biological Macromolecules*, 319, 145375.

Rony, M. S. M., Bristy, N. Y., Sarkar, P., Debnath, P. K., Shilpi, J. A., Uddin, S. J. (2025). Chitosan from Labeo rohita fish scales and its effects on ameliorating high-carbohydrate-high-fat diet-induced metabolic dysfunction by modulating mRNA expression of LDL-R in obese rats. *Pharmacological Research-Natural Products*, 7, 100269.

Salama, A., Hesemann, P. (2025). Guanylated chitosan derivatives for the adsorption of anionic dyes: Performance and mechanism. *International Journal of Biological Macromolecules*, 311, 143852.

Shi, Y., Wang, L., Wei, W., Ding, S., Zhang, W., Li, F., Li, Z., Meng, X. (2025). Chitosan sponge doped with high-yield microporous carbon nanospheres for rapid hemostasis. *International Journal of Biological Macromolecules*, 317, 144838.

Silva, F. M., de Camargo, R. V., Guedes, D. F. C., de Assis, H. C., Lopes-Olhê, F. C., da Cruz Filho, A. M. (2025). Enhancing bond strength and sealer penetration of resin-based Sealers: The impact of chitosan/genipin solution in root canal final irrigation. *International Journal of Adhesion and Adhesives*, 142, 104073.

Simon, A., Pramitasari, R., Anugrah, D. S. B., Waturangi, D. E. (2025). Development of intelligent packaging films utilising chitosan derived from black soldier fly exuviae and anthocyanin from red dragon fruit peel for monitoring banana ripeness. *International Journal of Biological Macromolecules*, 319, 145610.

Subramani, K., Tamilselvi Mohanasundaram, N. V., Srinivasan, S., Krishnaveni Selva Ganeshan, J., Incharoensakdi, A. (2025). Microalgae-based bio-fabrication using zinc oxide-chitosan nanocomposite for industrial effluent degradation and pollutant reduction. *International Journal of Biological Macromolecules*, 319, 145379.

Tan, X., Wu, X., Wang, X., Tao, Y., Li, Y., Tang, S., Zhang, Q., Sheng, Y. (2025). Chitosan-assisted dilute acid pretreatment with enhanced prehydrolysate fermentability and enzyme digestibility of pretreated substrates. *International Journal of Biological Macromolecules*, 310, 143231.

Tuesta, T., Castillo-Barzola, A., Linares, H., Ruiz-Pacco, G., Baena-Moncada, A. M., Valderrama-Negrón, A. C. (2025). Chitosan-based materials for food preservation: Enhancing shelf life and safety through sustainable nanoparticles and films. *Food Chemistry*, 486, 144589.

Vakili, M., Mojiri, A., Zwain, H. M., Yuan, J., Giwa, A. S., Wang, W., Gholami, F., Guo, X., Cagnetta, G., Yu, G. (2019). Effect of beading parameters on cross-linked chitosan adsorptive properties. *Reactive and Functional Polymers*, 144, 104354.

Vieira, R. M., Gauthier, E., Widanagamage, G., McKenzie, N., Dunn, K., Moghaddam, L., Halley, P., Brienzo, M. (2025). Uv-blocking and hydrophobicity improvement of chitosan-based film with lignin addition from a pilot scale-up Organosolve process of Banana Pseudostem. *International Journal of Biological Macromolecules*, 313, 144254.

Wagener, J., Wang, X., Becker, K. L., Aimanianda, V., Valsecchi, I., Gresnigt, M. S., Netea, M. G., Latge, J. P., Gow, N. A. R., van de Veerdonk, F. L. (2025). Immunomodulatory function of chitosan is dependent on complement receptor, 3. *The Cell Surface*, 14, 100146.

Wang, C., Song, Z., Ning, X., Wang, Y., Xiao, J. (2025a). Solid lipid–threshold guided chitosan coating of NLCs for improved gastrointestinal stability and curcumin bioavailability. *Colloids and Surfaces A: Physicochemical and Engineering Aspects*, 723, 137388.

Wang, J., Liu, Y., Li, X., Wu, S. (2025b). Peanut seed molds affected by chitosan loaded with different antimicrobial agents. *Journal of Stored Products Research*, 112, 102637.

Wang, S., Wang, Q., Xu, J., Yu, J., Wang, S. (2025c). In vitro fecal fermentation characteristics of starch-chitosan composite prepared by screw extrusion. Carbohydrate Polymers 361, 123657.

Xu, H., Ma, H., Guo, G., Wei, C., Zhang, X., Guo, M., Li, Y. (2025). Research progress on chitosan-based sutures: A review. *Carbohydrate Polymers*, 366, 123868.

Yakar, A., Türker, O. C., Çakmak, E., Yılmaz Baran, N., Baran, T. (2025). Fabrication of smart chitosan composite beads for alleviate boron toxicity in a model plant (*Lemna gibba*): Characterization, toxicity assessment, and boron removal. *Journal of Cleaner Production*, 519, 146038.

Yang, K., Dang, H., Liu, L., Hu, X., Li, X., Ma, Z., Wang, X., Ren, T. (2019). Effect of syringic acid incorporation on the physical, mechanical, structural and antibacterial properties of chitosan film for quail eggs preservation. *International Journal of Biological Macromolecules*, 141, 876-884.

Ye, L., Xu, Z., Li, Y., Wang, P., Yan, Y., Yan, J. (2025). A chitosan-integrated antibacterial protein composite nanocomplex derived from barnacle cement and spider silk. Journal of Biotechnology, 405, 57-71.

Yıldırım, M., Poyraz, S., Acet, Ö., Acet, B. Ö., Karakoç, V., Odabaşı, M. (2025). Chitosan hydrogels: Versatile platforms for drug delivery in cancer treatment, wound dressing, and 3D bioprinting applications. *International Journal of Biological Macromolecules*, 314, 144367.

Zacheski, J. B. A., Rojas, C. E. B., Gerhardt, R., Cappellari, N. P., de Farias, B. S., Ribeiro, A. C., Sant'Anna Cadaval Junior, T. R., de Almeida Pinto, L. A. (2025). Chitosan-based films enriched with anthocyanin compounds for potential use as food packaging. *International Journal of Biological Macromolecules*, 317, 144867.

Zeng, H., Wu, M., Gao, Z., Ma, Y., Zhou, H., Sun, J. (2025). Cross-linking oxidized rice straw and chitosan for preparing a fully bio-based adhesive with exceptional water-resistant bonding performance. *International Journal of Biological Macromolecules*, 315, 144416.

Zeraat pisheh, F., Pirnia, M., Falah, F., Tabatabaei-Yazdi, F., Mortazavi, S. A., Vasiee, A. (2025). Innovative edible coatings: Lactiplantibacillus plantarum postbiotics and chitosan as a clean-label strategy for sausage preservation. *Applied Food Research*, 5, 100958.

Zhang, G., Li, X., Xu, X., Tang, K., Vu, V. H., Gao, P., Chen, H., Xiong, Y. L., Sun, Q. (2019). Antimicrobial activities of irradiation-degraded chitosan fragments. *Food Bioscience*, 29, 94-101.

Zhang, H., Zhang, X., Wang, B., Ren, B., Yang, X. (2025). Synergistic multi-crosslinked networks in chitosan/Gallic acid/Fe^{3+} composite biofilms with high UV resistance and antioxidant properties. *International Journal of Biological Macromolecules*, 316, 144686.

Zhao, S., Wu, J., Li, M., Zhao, Y., Xiong, G., Wang, X., Peng, L. (2025). Assessment of chitosan nanoemulsion loading with perillaldehyde on the quality characteristics and microbial diversity of beef slices during refrigeration storage. *Food Chemistry*, 29, 102658.

Zhao, Y., Saldaña, M. D. A. (2019). Hydrolysis of cassava starch, chitosan and their mixtures in pressurized hot water media. *The Journal of Supercritical Fluids*, 147, 293-301

Zhou, Y. Q., Deng, X., Zhao, Z.-X., Wang, X. T., Jin, X. Q., Xiong, J. B., Chen, L. L., Guo, W.-H., Zhou, R.-B., Yin, D.-C. (2025). High-strength chitosan-sericin cryogel with synergistically reinforced networks for hemostasis. *International Journal of Biological Macromolecules*, 318, 145038.

Chapter 4

Targeted Antimicrobial Strategies Using Chitosan-Based Nanotechnologies: Mechanisms, Applications, and Future Directions

Duygu Aydemir[1,*]
and Nuriye Nuray Ulusu[2,3]

[1]Istanbul University, Institute of Child Health, Department of Pediatric Basic Sciences, Division of Medical Genetics, Istanbul, Türkiye
[2]Koç University, School of Medicine, Department of Medical Biochemistry, Sariyer, Istanbul, Türkiye
[3]Koç University Research Center for Translational Medicine (KUTTAM), Sariyer, Istanbul, Türkiye

Abstract

The global escalation of antimicrobial resistance (AMR) has emerged as a critical public health crisis, threatening the efficacy of existing antibiotics and the success of modern medical interventions. This chapter explores nanotechnology-driven approaches to combat AMR, with a particular focus on chitosan-based nanomaterials as multifunctional antimicrobial platforms. It provides an overview of the molecular mechanisms underlying antibiotic resistance, such as efflux pumps, target modification, enzymatic degradation, and biofilm formation. It delineates how nanostructured systems can circumvent these barriers. Chitosan, a naturally derived cationic polysaccharide, exhibits intrinsic antimicrobial activity and serves as an efficient nanocarrier for targeted,

* Corresponding Author's Email: duygu.aydemir@istanbul.edu.tr

In: Chitosan
Editor: Denis Hurst
ISBN: 979-8-90134-079-0

controlled, and synergistic drug delivery. Its unique physicochemical and biological properties enable enhanced permeability, disruption of bi, and pathogen-specific targeting while maintaining biocompatibility. The chapter also highlights diverse biomedical applications of chitosan nanocomposites, including antimicrobial wound dressings, infection-resistant medical device coatings, advanced drug delivery systems, and antifungal or antiviral formulations. Furthermore, emerging hybrid systems integrating chitosan with metallic nanoparticles, nanoflowers, and quantum dots are discussed as next-generation solutions for overcoming multidrug resistance. Finally, the chapter addresses translational challenges related to toxicity, large-scale production, and regulatory standardization, outlining future directions for the safe and sustainable deployment of chitosan-based nanotechnologies in antimicrobial therapy.

Keywords: antimicrobial resistance, chitosan, nanotechnology, biofilm, nanocarrier, quantum dots, wound healing, drug delivery

1. Introduction

Antimicrobial resistance (AMR) has evolved into one of the twenty-first century's most formidable public health challenges, posing a profound threat to global morbidity, mortality, and healthcare sustainability. Although the discovery and clinical deployment of antibiotics revolutionized modern medicine, enabling effective management of bacterial infections and facilitating surgical and oncological interventions, their therapeutic efficacy is now being critically undermined by the emergence and dissemination of multidrug-resistant (MDR) and extensively drug-resistant (XDR) pathogens [1]. The relentless misuse, overuse, and subtherapeutic administration of antibiotics across human medicine, veterinary practice, and agricultural sectors have imposed intense selective pressure on microbial populations, accelerating the evolution and horizontal transfer of resistance determinants. Consequently, infections caused by previously manageable bacterial species have become increasingly resistant to conventional antibiotic therapy, rendering once-routine clinical procedures riskier and more challenging to manage [1, 2].

Table 1. Detailed antimicrobial properties and mechanisms of chitosan and its nanostructured derivatives

Mechanism / Property	Molecular and Cellular Mechanism	Representative Microorganisms	Chitosan Form / Derivative	Biomedical and Technological Applications	Key Outcomes / Advantages
Electrostatic interaction with microbial membranes	The protonated amino groups ($-NH_3^+$) of chitosan bind to negatively charged microbial surfaces (LPS in Gram-negatives; teichoic acids in Gram-positives), increasing membrane permeability and causing leakage of intracellular contents.	Staphylococcus aureus, Escherichia coli, Pseudomonas aeruginosa, Klebsiella pneumoniae	Native chitosan, chitosan nanoparticles (CSNPs), chitosan–Ag nanocomposites	Wound dressings, implant coatings, antimicrobial films	Broad-spectrum bactericidal action; rapid membrane disruption; low resistance probability
Biofilm inhibition and disruption	Chitosan penetrates biofilm matrices, interferes with quorum-sensing signals (AI-2, acyl-homoserine lactones), and disrupts extracellular polymeric substance (EPS) structure, leading to biofilm dispersion.	P. aeruginosa, S. epidermidis, E. coli biofilms	Chitosan–metal hybrids (Ag–CS, ZnO–CS), chitosan–quantum dots	Anti-biofilm coatings, chronic wound therapy, dental materials	Prevents biofilm formation and recurrence; restores antibiotic susceptibility
Chelation of essential metal ions	The hydroxyl and amino groups of chitosan chelate divalent metal ions (Ca^{2+}, Mg^{2+}, Fe^{2+}), depriving microbes of cofactors required for enzymatic activity and cell wall stabilization.	Candida albicans, S. aureus, Bacillus subtilis	Low–molecular–weight chitosan (LMW–CS), modified chitosan oligosaccharides	Wound infection control, plant protection formulations	Inhibits enzymatic activity and spore germination; enhances bacteriostasis
DNA and protein synthesis inhibition	Chitosan or its oligomers penetrate bacterial/fungal cytoplasm and bind to DNA through electrostatic attraction, inhibiting mRNA transcription and protein translation.	E. coli, C. albicans, Aspergillus niger	Chitosan nanoparticles, CS–AMP conjugates	Antimicrobial coatings, antifungal sprays, topical formulations	Inhibits replication and protein synthesis at sublethal doses; multi-targeted activity

Table 1. (Continued)

Mechanism / Property	Molecular and Cellular Mechanism	Representative Microorganisms	Chitosan Form / Derivative	Biomedical and Technological Applications	Key Outcomes / Advantages
Reactive oxygen species (ROS) generation	Chitosan–metal nanocomposites (e.g., Ag, ZnO, Cu) catalyze ROS formation (•OH, $O_2^{\bullet-}$), leading to oxidative stress, lipid peroxidation, and protein denaturation in microbes.	MDR P. aeruginosa, MRSA, Enterococcus faecalis	CS–AgNPs, CS–ZnO nanoflowers, CS–quantum dots	Smart wound dressings, disinfectant coatings	Synergistic oxidative and electrostatic killing; sustained release; high stability
Efflux pump inhibition	Chitosan nanoparticles reduce expression or activity of bacterial efflux systems (e.g., AcrAB–TolC), thereby restoring intracellular antibiotic concentration.	E. coli, Acinetobacter baumannii	Chitosan–antibiotic nanoformulations	Drug delivery platforms, combination antimicrobial therapy	Enhances antibiotic retention; resensitizes resistant bacteria
Quorum-sensing interference	Chitosan disrupts microbial communication systems regulating virulence and biofilm formation by binding to signaling molecules or quorum-sensing receptors.	S. aureus, P. aeruginosa	CS–essential oil nanocomposites, CS–curcumin nanoparticles	Anti-virulence coatings, surface disinfectants	Reduces bacterial pathogenicity without selective pressure
Antifungal activity	Electrostatic binding to fungal membrane phospholipids increases permeability; chelation of ions inhibits mycelial growth and spore germination; DNA binding interferes with replication.	C. albicans, Botrytis cinerea, Fusarium oxysporum	CS hydrogels, CS/CMC microspheres, CS–citral composites	Vaginal gels, agricultural antifungals, food preservation	Inhibits hyphal growth; enhances stability of antifungal compounds
Antiviral activity	Positively charged chitosan interacts with negatively charged viral envelopes or capsid proteins, blocking viral adsorption and entry; sulfated or quaternized chitosan mimics heparan sulfate receptors to inhibit infection.	HSV-1, influenza virus, SARS-CoV-2	Sulfated chitosan, quaternized chitosan nanoparticles	Nasal sprays, vaccine adjuvants, antiviral coatings	Broad-spectrum antiviral defense; non-toxic; enhances immune response

Mechanism / Property	Molecular and Cellular Mechanism	Representative Microorganisms	Chitosan Form / Derivative	Biomedical and Technological Applications	Key Outcomes / Advantages
Controlled drug delivery and synergism	Chitosan acts as a mucoadhesive, pH-responsive carrier enhancing drug solubility, stability, and sustained release; co-delivery with antibiotics or metal NPs yields synergistic antimicrobial effects.	MDR bacterial and fungal strains	CS–antibiotic complexes, CS–Ag/ZnO hybrid nanoparticles	Drug delivery systems, wound healing scaffolds	Localized therapy, reduced systemic toxicity, enhanced drug bioavailability
Biocompatibility and tissue regeneration	Biodegradable, non-toxic polymer supports fibroblast proliferation, collagen synthesis, and angiogenesis during wound repair while preventing infection.	Host tissue (non-pathogenic)	Chitosan hydrogels, sponges, membranes	Wound dressings, tissue-engineering scaffolds	Promotes healing, controls infection, supports cell adhesion and regeneration

According to recent comprehensive analyses from the Global Research on Antimicrobial Resistance (GRAM) project, antimicrobial resistance has been responsible for more than one million deaths annually due to drug-resistant infections between 1990 and 2021 [3]. Epidemiological modeling further projects that, if current trajectories remain unmitigated, mortality directly attributable to AMR is expected to increase by approximately 70% by 2050, potentially reaching 1.9 million deaths per year [3, 4]. When infections in which AMR acts as a contributory factor are included, the global mortality burden could surpass 8 million deaths annually, thereby rivaling major non-communicable diseases as a leading cause of death worldwide. Notably, demographic analyses reveal a pronounced age-related disparity: the incidence of AMR-associated mortality has risen sharply among individuals aged 70 years and older, reflecting both cumulative antibiotic exposure and age-related immunosenescence, whereas mortality among children under five years has demonstrated a gradual decline over the past decades, likely attributable to expanded vaccination coverage, improved infection control, and better pediatric healthcare infrastructure. These findings collectively emphasize the urgent need for coordinated global interventions to prevent AMR from becoming a dominant cause of morbidity and mortality by mid-century. Contemporary strategies to mitigate AMR primarily focus on optimizing existing antimicrobial regimens and developing novel therapeutic entities capable of circumventing resistance mechanisms [5].

Conventional pharmacological approaches encompass combination antibiotic therapy, structural modification of existing drugs to evade enzymatic degradation, and the application of enzyme inhibitors, such as β-lactamase inhibitors, to restore or potentiate antibiotic efficacy [2]. In parallel, a range of non-traditional therapeutics, including antimicrobial peptides (AMPs), bacteriophage therapy, probiotics, and immunomodulatory agents, are being explored as adjuncts or alternatives to conventional antibiotics. Despite these advances, therapeutic effectiveness remains constrained by biofilm-associated tolerance, efflux pump overexpression, and the limited discovery pipeline for new antibiotic classes [4, 5]. Furthermore, the injudicious and excessive use of antibiotics in both clinical and agricultural contexts has accelerated the selection pressure driving resistance evolution. Compounding these challenges, the prolonged, resource-intensive, and economically disincentivized nature of antibiotic discovery and development has hindered the introduction of new agents into the clinical arsenal [6].

Nanotechnology offers innovative solutions to overcome the limitations of traditional antibiotics by enabling targeted, controlled, and efficient delivery of antimicrobial agents. Nanoparticles (NPs), including metallic (Ag, Au, ZnO, Cu), polymeric (e.g., PLGA, chitosan), lipid-based, and carbon-based nanostructures, exhibit multimodal antimicrobial mechanisms, such as disruption of bacterial membranes, generation of reactive oxygen species (ROS), and interference with DNA and protein synthesis [7]. NPs can also penetrate biofilms, deliver antibiotics directly to infection sites, and be functionalized with ligands for pathogen-specific targeting, thereby minimizing systemic toxicity and reducing the development of resistance [8]. Chitosan, a naturally derived cationic polysaccharide obtained through the deacetylation of chitin, demonstrates substantial potential in mitigating AMR owing to its intrinsic antimicrobial properties and exceptional versatility as a nanocarrier. Due to their excellent biocompatibility and positively charged surface characteristics, chitosan nanoparticles serve as highly effective nanocarriers that exhibit intrinsic antimicrobial properties and possess the potential to act synergistically with conventional antibiotic agents (Table 1) [9]. Despite promising results, challenges such as nanoparticle toxicity, stability, large-scale production, and regulatory approval remain barriers to clinical translation. Nevertheless, nanotechnology is a potential next-generation platform for antimicrobial therapy and resistance mitigation [10].

This chapter presents a comprehensive and critical examination of nanotechnology-driven strategies developed to counteract AMR, particularly emphasizing chitosan-based nanomaterials, nanoflowers, and quantum dots as emerging multifunctional platforms. The discussion encompasses the physicochemical characteristics, structure–function relationships, and molecular mechanisms underlying their antimicrobial activity, highlighting their ability to enhance drug delivery, facilitate targeted release, and enable integrated theranostic applications with chitosan. Furthermore, the chapter synthesizes recent technological advancements and experimental insights in nanomaterial engineering while addressing the key translational challenges associated with biocompatibility, toxicity, regulatory frameworks, and large-scale synthesis. Finally, it outlines future perspectives and research directions, underscoring the transformative potential of nanomaterial-based antimicrobial systems to redefine infection control paradigms, restore antibiotic efficacy, and contribute to the global effort against multidrug-resistant pathogens.

2. Mechanism of Antibiotic Resistance

The administration of antimicrobial agents in their free form is associated with several inherent limitations, including poor selectivity toward infection sites, particularly those associated with biofilms, and a high likelihood of premature degradation, metabolism, or systemic clearance before reaching the intended target. Moreover, due to the intrinsic tolerance of biofilms to conventional antibiotics, substantially elevated therapeutic doses are often required to achieve clinical efficacy. Such dosing regimens increase the risk of systemic toxicity when employing traditional delivery routes such as oral, inhalational, or parenteral administration. Consequently, adverse effects, including hypersensitivity reactions, hepatotoxicity, nephrotoxicity, and disruption of the commensal microbiota, have been frequently reported [11].

AMR represents a highly complex and multifactorial biological phenomenon that enables microorganisms to survive and proliferate despite exposure to antimicrobial agents with inhibitory or bactericidal activity. Resistance may be categorized as either intrinsic (natural), denoting an inherent characteristic of specific bacterial taxa, or acquired, which emerges through spontaneous genetic mutations or the horizontal transfer of resistance determinants. Both forms operate through sophisticated molecular mechanisms that impede the antimicrobial agent's access to its cellular target, modify or protect the target structure, or enzymatically inactivate the compound. The dynamic interplay among these mechanisms has culminated in the emergence and global dissemination of MDR and extensively drug-resistant (XDR) pathogens, posing a profound and escalating challenge to contemporary healthcare systems [12].

2.1. Reduced Membrane Permeability

Reduced membrane permeability constitutes a fundamental mechanism underlying antibiotic resistance in Gram-negative bacteria. The outer membrane (OM) serves as a selective permeability barrier, restricting the entry of antimicrobial agents, particularly those targeting intracellular processes. Resistance can emerge through alterations in both the lipid and protein components of the OM [13]. Structural remodeling of lipopolysaccharides (LPS), such as the incorporation of 4-aminoarabinose, phosphoethanolamine, or palmitate into lipid A, reduces the net negative charge and enhances molecular packing, thereby limiting the diffusion of cationic and hydrophobic antibiotics, including polymyxins and aminoglycosides. Simultaneously, changes in the expression or structure of porin proteins, such as OmpF and OmpC, impede the influx of hydrophilic antibiotics like β-lactams, tetracyclines, and fluoroquinolones [3]. Mutations that constrict the porin channel or regulatory mechanisms that downregulate porin synthesis can markedly decrease antibiotic permeability, often in concert with increased efflux activity. Collectively, these modifications result in a formidable outer membrane barrier that diminishes intracellular antibiotic accumulation, contributing substantially to multidrug resistance in pathogenic bacteria [14].

2.2. Increased Drug Efflux

Efflux pumps represent a critical determinant of bacterial multidrug resistance by serving as active barriers to intracellular antibiotic accumulation. These energy-dependent transport systems effectively extrude toxic compounds from the cytoplasm and periplasm, thereby maintaining sub-inhibitory intracellular concentrations of diverse antimicrobial agents [15]. Among the various efflux pump families, the resistance–nodulation–division (RND) superfamily, predominantly found in Gram-negative bacteria, and the major facilitator superfamily (MFS), common in Gram-positive organisms, are of clinical relevance. Prototypical RND systems, such as the AcrAB–TolC complex in Escherichia coli and the MexAB–OprM system in Pseudomonas aeruginosa, form tripartite assemblies that span the inner and outer membranes, enabling the direct extrusion of substrates into the external milieu. Structural analyses of AcrB have revealed multiple substrate-binding pockets and conformational cycling mechanisms that confer remarkable substrate promiscuity, allowing the efflux of chemically diverse compounds, including antibiotics, detergents, dyes, and biocides [16]. Overexpression of efflux pumps may arise from mutations in local repressors (e.g., *acrR*, *tetR*) or global transcriptional regulators (e.g., *marA*, *soxS*, *ramA*), as well as in response to environmental cues such as bile salts, indole, and iron limitation. The resulting phenotype is characterized by broad-spectrum resistance encompassing β-lactams, fluoroquinolones, macrolides, and tetracyclines. Notably, the simultaneous occurrence of efflux pump overexpression and decreased outer membrane permeability exerts a synergistic effect, collectively strengthening bacterial resilience against antibiotic stress and severely limiting therapeutic efficacy [17].

2.3. Target Modification by Mutation or Recombination

Many antibiotics exert their bactericidal or bacteriostatic effects through high-affinity interactions with specific molecular targets that are essential for bacterial survival. However, genetic alterations within the loci encoding these targets can markedly reduce antibiotic binding affinity while preserving the protein's physiological function, thereby conferring resistance [18]. One well-characterized example involves point mutations in the quinolone resistance–determining regions (QRDRs) of the *gyrA* and *parC* genes, which encode the A subunits of DNA gyrase and topoisomerase IV, respectively. These enzymes are critical for DNA replication and supercoiling, and amino acid substitutions, particularly at conserved residues such as Ser83 and Asp87 in GyrA or Ser80 in ParCn, significantly diminish fluoroquinolone binding without impairing catalytic activity, thereby conferring high-level resistance across diverse Gram-negative and Gram-positive species [19].

A similar paradigm operates in β-lactam resistance, where structural remodeling of penicillin-binding proteins (PBPs) reduces acylation by β-lactam antibiotics. In *Streptococcus pneumoniae* and *Neisseria gonorrhoeae*, homologous recombination events between related species result in mosaic PBPs that exhibit reduced β-lactam affinity while retaining transpeptidase function, thereby enabling continued peptidoglycan cross-linking in the presence of inhibitory drug concentrations [20]. In *Staphylococcus aureus*, the acquisition of the staphylococcal cassette chromosome mec (SCCmec), which carries the mecA gene, introduces a novel transpeptidase, PBP2a, characterized by a sterically constrained active site with a markedly diminished affinity for β-lactams. The expression of PBP2a enables cell wall synthesis to proceed even under high concentrations of methicillin and other β-lactams, thereby defining the methicillin-resistant Staphylococcus aureus (MRSA) phenotype. Collectively, these molecular adaptations illustrate a convergent evolutionary strategy whereby bacteria decouple antibiotic binding from target inhibition, preserving vital physiological functions while evading antimicrobial action, a central hallmark of target-mediated antibiotic resistance [21].

2.4. Target Protection and Chemical Modification

Target protection represents a sophisticated resistance mechanism wherein bacteria employ accessory proteins to sterically or allosterically shield antibiotic-binding sites, thereby preserving the activity of the target macromolecule. The Qnr family of pentapeptide repeat proteins exemplifies this strategy in the context of fluoroquinolone resistance. Qnr proteins associate with DNA gyrase and topoisomerase IV, stabilizing the enzyme–DNA complex and preventing quinolone-mediated stabilization of the cleavage complex that would otherwise result in lethal double-stranded DNA breaks [22]. This interaction effectively reduces the antibiotic's ability to inhibit DNA replication while maintaining the essential catalytic functions of the enzymes. A parallel mechanism operates in ribosomal protection against tetracyclines, mediated by the Tet(M) and Tet(O) proteins. These GTP-dependent translation factors bind to the ribosome and induce conformational changes that dislodge tetracycline from its binding site on the 30S subunit, thereby restoring aminoacyl-tRNA accommodation and resuming normal protein synthesis [23].

In contrast, target modification through post-translational chemical alteration directly alters the molecular architecture of the antibiotic binding site, thereby reducing or abolishing drug affinity. The erm (erythromycin ribosomal methylase) gene family encodes S-adenosylmethionine–dependent methyltransferases that methylate specific adenine residues, typically A2058, in domain V of 23S rRNA. This methylation sterically hinders the binding of macrolide, lincosamide, and streptogramin (MLS) antibiotics to the peptidyl transferase center, conferring the well-documented MLS_B resistance phenotype [24]. Similarly, the cfr (chloramphenicol–florfenicol resistance) gene encodes an rRNA methyltransferase that modifies A2503 in 23S rRNA, broadening resistance to multiple ribosome-targeting agents, including phenicols, oxazolidinones (such as linezolid), and pleuromutilins. Importantly, both *erm* and *cfr* genes are often plasmid-borne or associated with transposable elements, enabling their rapid horizontal dissemination among diverse bacterial taxa. Collectively, target protection and target modification exemplify two complementary molecular strategies that allow bacteria to preserve critical cellular processes while evading inhibition by structurally and mechanistically diverse antibiotics [25].

2.5. Alteration of Cell Envelope Targets

Antibiotics that disrupt bacterial envelope integrity, such as colistin and daptomycin, are increasingly confronted by target-based and structural resistance mechanisms that preserve membrane function while diminishing the efficacy of these drugs. In Gram-negative bacteria, resistance to colistin, a last-resort polymyxin antibiotic, frequently arises through mutations in the PmrAB and PhoPQ two-component regulatory systems [26]. These mutations result in constitutive activation of the *pmrC* gene (also known as *eptA*), which encodes a phosphoethanolamine transferase that covalently modifies the lipid A component of lipopolysaccharide (LPS). The addition of phosphoethanolamine decreases the net negative charge of the bacterial outer membrane, thereby reducing the electrostatic attraction between colistin's cationic peptide moieties and the anionic LPS surface. The consequence is a profound reduction in colistin binding affinity and a substantial increase in resistance, particularly among *Enterobacteriaceae* and *Pseudomonas* species [27].

In Gram-positive bacteria, resistance to daptomycin, a calcium-dependent lipopeptide that targets the cytoplasmic membrane, arises from alterations in membrane charge and composition, similar to those observed in resistance to other antibiotics. Mutations in the mprF gene, which encodes lysyl-phosphatidylglycerol synthase, lead to enhanced addition of L-lysine residues to phosphatidylglycerol molecules, decreasing the overall anionic surface charge of the membrane. This electrostatic modification repels the cationic daptomycin–calcium complex, thereby reducing its insertion and oligomerization within the membrane, a prerequisite for its bactericidal activity [28]. Beyond these biochemical modifications, structural reorganization of membrane domains provides an additional layer of resistance, particularly in enterococci. Resistant strains exhibit a redistribution of cardiolipin-enriched microdomains, which displaces daptomycin-binding sites away from the division septum, the primary locus of its antimicrobial action. This spatial remodeling prevents the antibiotic from accessing its target regions, effectively neutralizing its bactericidal effect without altering membrane composition or charge. Collectively, these adaptive strategies encompassing lipid modification, charge alteration, and membrane domain reorganization highlight the remarkable plasticity of bacterial envelope physiology in countering envelope-active antibiotics [29].

2.6. Hydrolytic Enzymes

Among the diverse mechanisms of antibiotic resistance, enzymatic hydrolysis represents one of the most pervasive and clinically consequential strategies employed by bacteria to neutralize antimicrobial agents. The canonical example is the β-lactamase, first described in 1940, which catalyzes the hydrolytic cleavage of the β-lactam ring, the structural motif essential for the bactericidal activity of β-lactam antibiotics. This reaction renders the antibiotic molecule incapable of acylating penicillin-binding proteins (PBPs), thereby abolishing its ability to inhibit bacterial cell wall synthesis. Over the subsequent decades, extensive genetic diversification and horizontal gene transfer have led to a wide array of β-lactamase families exhibiting distinct substrate profiles and inhibitor susceptibilities [30].

Of clinical importance are the extended-spectrum β-lactamases (ESBLs), which evolved through point mutations in classical TEM and SHV enzymes, expanding their hydrolytic spectrum to include third-generation cephalosporins and monobactams. Even more alarming are the carbapenemases, such as KPC (Klebsiella pneumoniae carbapenemase), NDM (New Delhi metallo-β-lactamase), VIM (Verona integron-encoded metallo-β-lactamase), and OXA-48–type enzymes, which collectively confer resistance to nearly all β-lactam antibiotics, including the last-line carbapenems [31]. These enzymes differ mechanistically: serine carbapenemases (e.g., KPC, OXA-48) utilize an active-site serine residue for nucleophilic attack, whereas metallo-β-lactamases (e.g., NDM, VIM) require zinc ions for catalytic activity.

The genetic mobility of β-lactamase genes has been central to their global dissemination. Frequently located on conjugative plasmids, transposons, or integrons, these genes are readily transmitted among Enterobacteriaceae, Pseudomonas aeruginosa, and Acinetobacter baumannii, promoting rapid interspecies spread within hospital and community environments [32]. A particularly successful lineage, the CTX-M family, derived from environmental *Kluyvera* species, has achieved pandemic distribution, with CTX-M-15 emerging as the most prevalent ESBL worldwide. The proliferation of such enzymes exemplifies the dynamic interplay between microbial evolution, genetic mobility, and the use of clinical antibiotics, underscoring the urgent need for novel inhibitors and robust antimicrobial stewardship strategies to mitigate the escalating threat of β-lactamase-mediated resistance [33].

2.7. Chemical Modification of Antibiotics

Bacterial resistance to aminoglycosides is frequently mediated by enzymatic modification of the antibiotic molecule, a mechanism that effectively abolishes its ability to bind the ribosomal target. These resistance enzymes catalyze covalent modifications, most commonly acetylation, phosphorylation, or adenylation of specific hydroxyl or amino groups within the aminoglycoside scaffold. Such chemical alterations induce steric hindrance and electrostatic repulsion, preventing the antibiotic from associating with the 16S rRNA of the 30S ribosomal subunit, which is essential for its bactericidal activity [34]. Three principal families of aminoglycoside-modifying enzymes (AMEs) have been characterized: aminoglycoside acetyltransferases (AACs), aminoglycoside phosphotransferases (APHs), and aminoglycoside nucleotidyltransferases (ANTs). Each class exhibits distinct catalytic mechanisms and substrate specificities. AACs utilize acetyl-CoA as a donor to transfer acetyl groups to amino residues. In contrast, APHs and ANTs employ ATP or GTP to catalyze phosphorylation or adenylation of hydroxyl groups, respectively. Collectively, these enzymes confer high-level and often cross-class resistance, as even a single modification can abolish the binding affinity of multiple aminoglycosides, including gentamicin, kanamycin, tobramycin, and amikacin [35].

Recent comparative genomic and metagenomic investigations have uncovered genetic clusters encoding multiple AMEs within both chromosomal and plasmid contexts, notably in *Campylobacter*, *Enterococcus*, *Pseudomonas*, and *Acinetobacter* species. This genetic redundancy enhances bacterial adaptability and persistence in the face of antimicrobial pressure. Moreover, the frequent localization of AME genes within mobile genetic elements, such as transposons, integrons, and conjugative plasmids, facilitates horizontal gene transfer (HGT) across taxonomically diverse bacteria, thereby accelerating the global dissemination of aminoglycoside resistance. These findings underscore the evolutionary plasticity and ecological mobility of AME-encoding loci, which collectively pose a formidable challenge to the therapeutic utility of aminoglycosides in clinical settings [36].

2.8. Biofilm

Biofilm represents highly organized, sessile microbial communities embedded within a self-produced extracellular polymeric substance (EPS) matrix adhering to biotic or abiotic surfaces. This complex three-dimensional structure provides an exceptionally protective environment, enabling microorganisms to survive under adverse physiological and environmental conditions, such as desiccation, nutrient limitation, and host immune responses. Biofilm formation is now recognized as a critical factor in the pathogenesis of chronic and recurrent infections, particularly in immunocompromised individuals and patients with indwelling medical devices, including catheters, prosthetic heart valves, pacemakers, contact lenses, and joint implants. Epidemiological data indicate that approximately half of all nosocomial infections in immunodeficient patients are attributable to biofilm-forming pathogens [37].

The biofilm mode of growth confers a distinct survival advantage, primarily by enhancing microbial tolerance to antimicrobial agents. Cells residing within biofilms display MICs and minimum bactericidal concentrations (MBCs) compared with their planktonic counterparts. While certain antibiotics, such as colistin or imipenem, may transiently reduce bacterial viability within biofilms, complete eradication generally requires concentrations that are clinically unachievable and often associated with significant systemic toxicity [38]. Consequently, biofilm-associated infections are notoriously refractory to conventional antibiotic therapy and frequently necessitate device removal or surgical intervention. Several interdependent mechanisms contribute to the elevated antimicrobial resistance observed in bacteria associated with biofilms. First, the EPS matrix functions as a physical and chemical barrier, restricting the diffusion of antibiotics and inactivating agents through adsorption or enzymatic degradation. Second, the biofilm microenvironment is characterized by pronounced gradients of oxygen, nutrients, and pH, which create physiologically heterogeneous subpopulations. Bacteria in deeper, anoxic regions often enter a slow-growing or dormant state, rendering them inherently less susceptible to antibiotics that target active cellular processes, such as β-lactams and aminoglycosides [39].

Third, biofilms harbor persister cells, a small fraction of metabolically quiescent bacteria capable of surviving lethal concentrations of antibiotics without acquiring genetic resistance. These persisters can repopulate the biofilm after treatment cessation, thereby perpetuating chronic infection. Fourth, the close cellular proximity within biofilms facilitates horizontal gene transfer via plasmids and transposons, promoting the dissemination of resistance determinants. Additionally, exposure to subinhibitory antibiotic concentrations has been shown to increase mutation rates, further accelerating the evolution of resistance. Another critical factor is the protection from host immune defenses afforded by the biofilm matrix [40].

The EPS layer impedes the access and activity of immune components, including phagocytes and complement proteins, thereby attenuating host-mediated bacterial clearance. This immune evasion contributes significantly to the persistence and recurrence of biofilm-related infections [41]. Species-specific characteristics also influence biofilm-associated resistance. For instance, the mucoid phenotype of *Pseudomonas aeruginosa*, prevalent in cystic fibrosis patients, produces an alginate-rich matrix that limits antibiotic penetration and oxygen diffusion, conferring marked tolerance to aminoglycosides such as tobramycin. Similarly, biofilms formed by *Staphylococcus aureus* or *Escherichia coli* demonstrate increased survival in hypoxic and nutrient-depleted environments. Importantly, recent findings indicate that biofilm-like aggregates do not need to be surface-attached to exhibit resistance [42].

Non-surface-associated microbial clusters found in chronic wounds, otitis media, and cystic fibrosis sputum display comparable antibiotic tolerance and immune evasion capabilities. Collectively, these multifactorial mechanisms render biofilms one of the most formidable barriers to effective antimicrobial therapy. The persistence of biofilm-associated infections underscores the need for innovative therapeutic interventions [43]. Contemporary strategies under investigation include physical removal or replacement of contaminated devices, disruption of cyclic di-GMP signaling, inhibition of quorum-sensing pathways, and the development of novel anti-biofilm agents capable of dispersing established microbial communities. Emerging approaches such as nanotechnology-based drug delivery systems, enzyme-mediated biofilm degradation, and immune-modulatory therapeutics offer promising avenues to overcome the inherent resistance and resilience of biofilm-associated pathogens [44].

3. Nanotechnology-Based Strategies for Combating Antimicrobial Resistance

NPs typically range in diameter from 1 to 100 nm and possess distinctive physicochemical and biological characteristics, including exceptionally high surface-to-volume ratios, tunable surface chemistry, and the capacity to traverse complex biological barriers. These unique attributes render NPs powerful antimicrobial agents as well as highly efficient drug-delivery platforms. Metallic nanoparticles, such as silver (AgNPs), zinc oxide (ZnO), copper (CuNPs), and titanium dioxide (TiO_2), exhibit broad-spectrum bactericidal activity mediated through multiple mechanisms, including the generation of reactive oxygen species (ROS), disruption of cellular membrane integrity, and interference with vital biomolecular and enzymatic processes [45]. Conversely, polymeric nanoparticles (e.g., chitosan, poly(lactic-co-glycolic acid) [PLGA]) and lipid-based nanocarriers, such as liposomes and solid lipid nanoparticles (SLNs), offer precise, localized, and sustained antibiotic release, thereby enhancing drug accumulation at infection sites while minimizing systemic toxicity. Similarly, carbon-based nanostructures, including graphene oxide (GO) and carbon nanotubes (CNTs), induce both physical and oxidative damage to bacterial cell membranes, effectively inhibiting biofilm development and microbial colonization [46].

Nanotechnology further enables targeted antimicrobial delivery, achieved through the functionalization of nanoparticles with specific ligands, antibodies, or peptides that can selectively bind to bacterial surface receptors or virulence determinants. This targeted approach enhances therapeutic precision, reduces off-target effects, and preserves the integrity of commensal microbiota. Moreover, stimuli-responsive nanocarriers, engineered to respond to environmental triggers such as pH, redox potential, or temperature, facilitate spatiotemporally controlled release of therapeutics at the site of infection, thus optimizing pharmacodynamic efficacy [47]. Beyond therapeutic applications, nanotechnology makes substantial contributions to the development of nanoparticle-based vaccines, which enhance antigen stability, facilitate efficient antigen presentation, and elicit robust immune responses against antibiotic-resistant pathogens, including *Staphylococcus aureus* and *Pseudomonas aeruginosa*. In parallel, AI- and ML-integrated nano-sensing platforms are being developed for the rapid detection of pathogenic species and resistance determinants, enabling real-time diagnostics and informed antimicrobial stewardship. Despite these promising advances, several critical challenges impede the clinical translation of nanotechnology-based antimicrobial systems [48].

Paramount among these are nanoparticle-associated cytotoxicity, concerns regarding biodegradability, and the lack of harmonized regulatory frameworks governing their biomedical applications. Furthermore, the scalability and cost-effectiveness of nanoparticle synthesis remain limiting factors. A comprehensive assessment of the long-term biological interactions and environmental implications of nanomaterials is also imperative to ensure their safe and sustainable integration into therapeutic practice [49]. To enhance the therapeutic efficacy and safety profile of antimicrobial therapy, advanced drug-delivery systems (DDSs) have been developed to enable controlled, localized, and sustained release of therapeutic compounds at infection sites, maintaining drug concentrations within the optimal therapeutic window. Extensive research efforts have therefore been directed toward improving the performance of existing antimicrobial agents against biofilm-associated infections by encapsulating them into particulate delivery platforms, including nanoparticles (NPs, <1 μm) and microparticles (MPs,>1 μm). These particulate systems offer multiple advantages, including the capacity for targeted or stimulus-triggered drug release, protection of the encapsulated agent from enzymatic or chemical degradation, and the flexibility to incorporate both hydrophilic and lipophilic therapeutics within a single carrier framework [50].

4. Antimicrobial Effects and Applications of Chitosan

Chitosan, a naturally derived cationic polysaccharide obtained through the deacetylation of chitin, demonstrates substantial potential in mitigating AMR owing to its intrinsic antimicrobial properties and exceptional versatility as a nanocarrier. The polymers' positively charged amino groups interact electrostatically with the negatively charged components of bacterial cell envelopes, leading to pronounced perturbation of membrane integrity, leakage of intracellular constituents, and subsequent inhibition of nucleic acid and protein synthesis. These broad-spectrum, multi-targeted interactions substantially diminish the likelihood of bacteria developing stable resistance mechanisms. When engineered into nanoparticulate formulations, chitosan markedly enhances the therapeutic performance of conventional antibiotics by improving their physicochemical stability, membrane permeability, and enabling sustained or stimuli-responsive release profiles [51].

Chitosan nanoparticles are further capable of penetrating and destabilizing microbial biofilms, suppressing quorum-sensing signaling pathways, and attenuating efflux pump activity, collectively restoring antibiotic susceptibility in MDR bacterial strains. Notably, they exhibit pronounced efficacy against Gram-negative pathogens through electrostatic permeabilization of the outer membrane and chelation-induced destabilization of lipopolysaccharide structures. In addition to their antimicrobial and adjuvant properties, the inherent biocompatibility, biodegradability, and non-toxic nature of chitosan render it particularly suitable for diverse biomedical applications, including wound healing, implant coatings, and localized infection management. Overall, chitosan and its nanostructured derivatives constitute a promising and multifunctional nanotechnological platform for combating antimicrobial resistance through the integration of intrinsic antimicrobial activity, targeted and controlled drug delivery, and biofilm disruption, while simultaneously minimizing cytotoxicity and the emergence of new resistance phenotypes [52].

4.1. Wound Dressing

Wound healing is a multifaceted physiological process encompassing hemostasis, inflammation, proliferation, and tissue remodeling. The presence of microbial infection, particularly by MDR pathogens such as *Staphylococcus aureus*, *Pseudomonas aeruginosa*, and *Klebsiella pneumoniae*, significantly delays this process and increases the risk of chronic wounds. Consequently, the development of biocompatible, biodegradable, and antimicrobial wound dressings has become a significant focus in regenerative medicine. Among natural polymers, chitosan, a deacetylated derivative of chitin, has emerged as a leading material due to its intrinsic antimicrobial, hemostatic, and wound-healing properties [53]. Chitosan-based dressings can be fabricated into various forms, including hydrogels, sponges, membranes, nanofibers, and films, each providing unique structural and functional advantages. These systems not only create a moist and protective healing environment but also act as delivery platforms for antimicrobial and bioactive agents, thereby enhancing tissue regeneration and combating infection. The antimicrobial activity of chitosan in wound environments primarily arises from its polycationic character, which facilitates electrostatic binding to the negatively charged membranes of microbes, leading to cell wall disruption, cytoplasmic leakage, and bacterial death [54].

Additionally, low–molecular–weight chitosan or chitosan nanoparticles can penetrate bacterial cells, bind to DNA, and inhibit transcription and protein synthesis. Chitosan also acts as a metal-chelating agent, depriving bacteria of essential ions such as Ca^{2+} and Mg^{2+}, while forming a protective film over the wound surface that limits nutrient access and bacterial colonization. From a wound-healing perspective, chitosan promotes fibroblast proliferation, collagen synthesis, and angiogenesis, while simultaneously exhibiting anti-inflammatory and hemostatic effects. These combined properties enable chitosan-based dressings to maintain a sterile, hydrated, and biocompatible microenvironment conducive to tissue repair [55].

Chitosan hydrogels possess a high water-retention capacity and are ideal for maintaining a moist wound interface, thereby accelerating epithelialization. Their porous and elastic structure allows gas exchange while absorbing exudate. Functionalization with antibiotics (e.g., sulfadiazine, levofloxacin) or nanoparticles (e.g., Ag, Cu, ZnO) enhances their antimicrobial performance [56]. For example, chitosan–silver nanoparticle (CS–AgNP) hydrogels have demonstrated potent activity against MRSA and P. aeruginosa, while maintaining cytocompatibility for fibroblasts and keratinocytes. Chitosan-based sponges are three-dimensional foams with high porosity and swelling ability, providing mechanical cushioning and exudate absorption. They are handy for treating deep or exudative wounds. Incorporation of hyaluronic acid, Aloe vera, or curcumin enhances their regenerative and antimicrobial properties. In diabetic wound models, chitosan–Ag nanoparticle sponges eradicated resistant bacteria and improved wound closure rates compared to control treatments [57].

Several commercial chitosan-based wound dressings are currently available, including HidroKi, Chitopack, Tegasorb, and KytoCel, which utilize native or modified chitosan matrices. These products have demonstrated clinical efficacy in accelerating wound closure, controlling infection, and reducing exudate accumulation [58]. Ongoing innovations involve the integration of metallic nanoparticles, growth factors, and stimuli-responsive components to develop "smart" dressings capable of real-time infection sensing and on-demand drug release. Future research aims to optimize the biomechanical and biological performance of chitosan-based wound dressings through nanocomposite engineering, surface functionalization, and incorporation of bio-responsive sensors. Addressing standardization in antimicrobial assays, long-term biocompatibility, and scalable manufacturing remains critical for clinical translation. The convergence of nanotechnology, materials science, and tissue engineering positions chitosan as a cornerstone biopolymer for next-generation antimicrobial wound care systems [59].

4.2. Anti-Microbial Coatings for Medical Devices

Medical device–associated infections pose a significant clinical challenge, accounting for a substantial proportion of hospital-acquired infections (HAIs). Devices such as catheters, orthopedic implants, prosthetic valves, and wound dressings provide surfaces conducive to microbial colonization and biofilm formation, which confer enhanced tolerance to antibiotics and host immune defenses. To mitigate these risks, the development of biocompatible, antimicrobial coatings has become a key strategy in biomedical engineering. Among the various candidates, chitosan, a naturally derived polysaccharide, has emerged as an auspicious material due to its intrinsic antimicrobial activity, film-forming ability, biocompatibility, and biodegradability [60].

The antimicrobial effects of chitosan in medical device coatings primarily arise from its cationic nature, which enables electrostatic interactions with the negatively charged cell walls of microbial cells. This interaction leads to membrane destabilization, leakage of intracellular components, and cell death. Additionally, chitosan films can form a physical barrier on device surfaces, thereby preventing bacterial adhesion and the establishment of biofilms. At the nanoscale, chitosan nanoparticles or chitosan–metal composites can penetrate microbial membranes, bind to DNA, and inhibit transcription and protein synthesis, thereby exerting bactericidal activity [61]. Moreover, the metal-chelating properties of chitosan deprive microorganisms of essential ions such as Ca^{2+} and Mg^{2+}, which are necessary for cell wall integrity and enzymatic activity. Chitosan can be deposited onto various biomaterial surfaces, including titanium, stainless steel, silicone, polyurethane, and polyethylene, using techniques such as dip coating, layer-by-layer assembly, electrostatic deposition, or plasma grafting. These coatings enhance surface hydrophilicity and reduce bacterial adhesion while maintaining compatibility with mammalian cells [62].

The integration of silver nanoparticles (AgNPs) into chitosan matrices yields synergistic antimicrobial activity against *Staphylococcus aureus*, *Pseudomonas aeruginosa*, and *Escherichia coli*. The chitosan matrix stabilizes silver nanoparticles, controls the release of Ag^{+} ions, and reduces cytotoxicity, ensuring sustained antimicrobial efficacy on catheter or implant surfaces. These composites exhibit broad-spectrum activity through the generation of reactive oxygen species (ROS) and membrane oxidation. ZnO–chitosan coatings have shown effective inhibition of biofilm formation on titanium and silicone substrates used in dental and orthopedic implants. Moreover, chitosan can serve as a controlled-release carrier for antibiotics (e.g., vancomycin, gentamicin, ciprofloxacin) or antimicrobial peptides (e.g., LL-37, nisin), providing localized, prolonged drug release at the implant interface [63].

Such hybrid coatings minimize systemic toxicity while maintaining local antimicrobial concentrations above the minimum inhibitory level. Recent developments incorporate graphene oxide, polylactic acid (PLA), or polyethylene glycol (PEG) into chitosan matrices to improve mechanical durability, hydrophilicity, and antimicrobial resistance. These hybrid materials exhibit enhanced anti-adhesive properties and support fibroblast attachment, making them suitable for long-term, implantable devices [64]. Biofilm formation on medical devices remains a principal cause of persistent infection and antibiotic failure. Chitosan-based coatings disrupt the initial adhesion phase by altering surface charge and wettability, thereby preventing bacterial anchoring. Furthermore, chitosan's film-forming and swelling behavior limits nutrient availability and oxygen access to microbes, suppressing biofilm maturation. Functionalization with nanoparticles or quorum-sensing inhibitors (e.g., furanones, cinnamaldehyde) further enhances biofilm suppression, particularly against *S. epidermidis* and *P. aeruginosa* [65].

A significant advantage of chitosan over synthetic antimicrobials is its excellent biocompatibility and non-cytotoxicity. It promotes fibroblast proliferation and tissue integration, which are crucial for device acceptance and healing. Several commercial prototypes, such as chitosan-coated catheters and wound-contact layers, have demonstrated reduced infection rates and faster tissue recovery in preclinical and clinical studies [54]. Moreover, chitosan coatings can be enzymatically degraded *in vivo,* avoiding long-term accumulation or inflammatory reactions. Despite its promising antimicrobial performance, challenges remain in achieving uniform and durable coatings, as well as controlling the long-term release of antimicrobials without compromising mechanical stability. Future research focuses on stimuli-responsive coatings that release antimicrobials in response to pH or enzyme changes triggered by infection. Multifunctional hybrid coatings combining chitosan with bioactive ceramics, peptides, or photothermal agents. Standardized evaluation protocols to assess antimicrobial efficacy, cytocompatibility, and long-term stability under physiological conditions. The integration of chitosan-based technologies with nanomaterials, 3D printing, and intelligent sensor systems holds great promise for next-generation infection-resistant medical devices that actively prevent biofilm formation and support tissue regeneration [66].

4.3. Drug Delivery Systems

Chitosan has emerged as a versatile biomaterial in the design of drug delivery systems (DDS) due to its biocompatibility, biodegradability, non-immunogenicity, and intrinsic antimicrobial properties. Its cationic nature, conferred by protonated amino groups, facilitates mucoadhesion, controlled drug release, and targeted delivery to infection or inflammation sites. These properties make chitosan particularly suitable for the development of antimicrobial and anti-inflammatory delivery platforms that aim to overcome the limitations of conventional drug administration, such as poor solubility, short half-lives, and systemic toxicity [67]. The effectiveness of chitosan as a drug carrier stems from its unique combination of physicochemical and biological properties that collectively enhance drug stability, bioavailability, and therapeutic performance. Its mucoadhesive capacity arises from the protonated amino groups of chitosan, which electrostatically interact with negatively charged residues, such as sialic acid and sulfate groups, present on mucosal surfaces, including those of the nasal, intestinal, ocular, and pulmonary epithelia. This interaction prolongs the residence time of drug formulations at the absorption site, thereby increasing local drug concentration and improving absorption efficiency [68].

Furthermore, chitosan functions as a biological permeation enhancer by transiently opening tight junctions between epithelial cells, facilitating paracellular transport of hydrophilic molecules, peptides, and other macromolecular therapeutics. This reversible modulation of tight junction proteins, such as zonula occludens (ZO-1 and ZO-2), enhances permeability without causing structural damage to the epithelium [69]. In addition, chitosan's chemical versatility allows its modification through crosslinking or grafting to achieve controlled and sustained drug release. Such modifications impart pH-responsive, enzymatically degradable, or redox-sensitive behavior, enabling site-specific and prolonged drug delivery, particularly useful in infection sites characterized by acidic and enzyme-rich microenvironments. Beyond its carrier function, chitosan exhibits intrinsic antimicrobial activity that synergizes with co-delivered antibiotics or natural bioactives, enhancing bactericidal efficacy while reducing required drug dosages and mitigating the risk of antimicrobial resistance development. This unique combination of mucoadhesion, permeability enhancement, controlled release, and antimicrobial effects underscores chitosan's versatility as a next-generation platform for localized and systemic drug delivery applications [70].

4.4. Anti-Fungal and Anti-Viral Applications

Chitosan and its derivatives exhibit notable antifungal and antiviral properties, making them promising candidates for biomedical and agricultural applications. Their antifungal efficacy primarily arises from electrostatic interactions between the protonated amino groups of chitosan and the negatively charged fungal cell membranes, which increase membrane permeability, disrupt ionic balance, and lead to the leakage of intracellular constituents. In addition, chitosan can chelate essential metal ions required for fungal enzymatic activity and spore germination, thereby inhibiting mycelial growth and reproduction [71]. Recent formulations developed through spray-congealing techniques have further improved chitosan's performance as an antifungal carrier. For example, seven distinct microsphere formulations incorporating chitosan, sodium carboxymethylcellulose, poloxamers (Lutrol F68 and F127), and Gelucire 53/10 demonstrated high yields (>90%) and enhanced solubility and bioavailability. These poloxamer/Gelucire-based microspheres exhibited significant inhibitory activity against *Candida albicans* ATCC 10231, highlighting their potential for localized antifungal therapy, particularly in treating vaginal candidiasis with reduced dosing frequency [72].

Moreover, citral, an acyclic monoterpene aldehyde known to alter fungal mitochondrial structure, has shown synergistic activity when incorporated into chitosan/carboxymethylcellulose (CS/CMC) hydrogel microspheres. This combination effectively eradicated *Botrytis cinerea*, an invasive plant fungus, suggesting potential applications in plant protection and post-harvest preservation. Chitosan-based carriers thus not only stabilize chemically labile antifungal compounds, such as citral, but also extend their bioactivity and environmental stability [73].

Beyond antifungal activity, chitosan also demonstrates broad-spectrum antiviral potential, particularly against enveloped viruses such as herpes simplex virus (HSV), influenza, and coronaviruses. The antiviral mechanism is attributed to electrostatic binding between chitosan's positively charged amino groups and the negatively charged viral envelope or capsid proteins, which impedes viral adsorption and entry into host cells. Modified chitosan derivatives, such as sulfated, quaternized, and carboxymethylated chitosan, exhibit enhanced antiviral efficacy by mimicking heparan sulfate structures that compete with viral binding sites on host membranes. These properties make chitosan-based systems particularly suitable for the development of topical antiviral formulations, nasal sprays, and vaccine adjuvants, offering safe and biodegradable alternatives to conventional antiviral agents [74].

In summary, the multifunctional antimicrobial profile of chitosan, encompassing both antifungal and antiviral activities, arises from its structural versatility, biocompatibility, and ability to synergize with other bioactive compounds. Its application across medical, pharmaceutical, and agricultural domains underscores the value of chitosan as a biopolymeric platform for infection control and prevention.

Conclusion

Antimicrobial resistance (AMR) remains a mounting global health crisis, undermining the efficacy of existing antibiotics and threatening the success of modern healthcare interventions. Nanotechnology offers a transformative framework for overcoming the limitations of conventional antimicrobial therapies by enabling precise targeting, controlled drug release, and multimodal antimicrobial actions. Among these emerging systems, chitosan-based nanomaterials represent a compelling platform due to their intrinsic antimicrobial activity, biocompatibility, biodegradability, and tunable physicochemical characteristics. Through mechanisms such as membrane disruption, inhibition of biofilm formation, and suppression of efflux pump activity, chitosan-based nanostructures can significantly enhance the potency and durability of antimicrobial agents while minimizing systemic toxicity and the development of resistance.

Chitosan's versatility is further demonstrated in its wide-ranging biomedical applications, including antimicrobial wound dressings, anti-biofilm coatings for medical devices, advanced drug delivery systems, and antifungal or antiviral formulations. The combination of chitosan with metallic nanoparticles, nanoflowers, and quantum dots introduces multifunctional hybrid systems that merge antimicrobial, sensing, and regenerative capabilities within a single platform. Collectively, these advances position chitosan-based nanotechnologies as a cornerstone in the next generation of antimicrobial strategies, capable of addressing the complexity of multidrug-resistant infections through synergistic, targeted, and sustainable approaches.

Future Perspectives

While the potential of chitosan-based nanotechnologies is undeniable, translating these systems from laboratory research to clinical application requires overcoming several critical barriers. Future work should focus on optimizing synthesis protocols to achieve consistent particle size, stability, and reproducibility across production scales. Comprehensive *in vitro* and *in vivo* evaluations are crucial for fully elucidating pharmacokinetic behavior, biodegradation pathways, and long-term safety profiles. Furthermore, the establishment of standardized regulatory guidelines and interdisciplinary collaboration among material scientists, microbiologists, clinicians, and policymakers will be crucial to ensure safe clinical deployment.

The integration of chitosan-based nanomaterials with emerging technologies—such as artificial intelligence–assisted drug design, innovative biosensors, and stimuli-responsive materials—offers exciting opportunities for the development of intelligent, adaptive antimicrobial systems capable of real-time infection detection and on-demand therapeutic release. As the global community continues to face the escalating threat of AMR, chitosan-based nanotechnologies are poised to play a pivotal role in shaping the future of infection control and antimicrobial innovation.

References

[1] Muteeb G, Rehman MT, Shahwan M, Aatif M: Origin of Antibiotics and Antibiotic Resistance, and Their Impacts on Drug Development: A Narrative Review. *Pharmaceuticals*. 2023, 16:1615. 10.3390/ph16111615.

[2] Srivastava J, Chandra H, Nautiyal AR, Kalra SJS: Antimicrobial resistance (AMR) and plant-derived antimicrobials (PDAms) as an alternative drug line to control infections. *3 Biotech*. 2014, 4:451–60. 10.1007/s13205-013-0180-y.

[3] EClinicalMedicine: Antimicrobial resistance: a top ten global public health threat. *EClinicalMedicine*. 2021, 41:101221. 10.1016/j.eclinm.2021.101221.

[4] Zhang J, Liu M, Guo H, Gao S, Hu Y, Zeng G, Yang D: Nanotechnology-driven strategies to enhance the treatment of drug-resistant bacterial infections. *WIREs Nanomedicine and Nanobiotechnology*. 2024, 16:. 10.1002/wnan.1968.

[5] Munita JM, Arias CA: Mechanisms of Antibiotic Resistance. *Microbiol Spectr*. 2016, 4:. 10.1128/microbiolspec.VMBF-0016-2015.

[6] Sionov RV, Steinberg D: Targeting the Holy Triangle of Quorum Sensing, Biofilm Formation, and Antibiotic Resistance in Pathogenic Bacteria. *Microorganisms*. 2022, 10:1239. 10.3390/microorganisms10061239.

[7] Bharti S, Mukherji S, Mukherji S: Antiviral application of colloidal and immobilized silver nanoparticles. *Nanotechnology*. 2021, 32:205102. 10.1088/1361-6528/abe 489.

[8] Guerrero Correa M, Martínez FB, Vidal CP, Streitt C, Escrig J, de Dicastillo CL: Antimicrobial metal-based nanoparticles: a review on their synthesis, types and antimicrobial action. *Beilstein Journal of Nanotechnology*. 2020, 11:1450–69. 10.3762/bjnano.11.129.

[9] Aydemir D, Öztürk K, Arslan FB, Çalis S, Ulusu NN: Gemcitabine-loaded chitosan nanoparticles enhanced apoptotic and ferroptotic response of gemcitabine treatment

alone in the pancreatic cancer cells in vitro. *Naunyn Schmiedebergs Arch Pharmacol.* Published Online First: 17 June 2024. 10.1007/s00210-024-03193-6.

[10] Poon W, Kingston BR, Ouyang B, Ngo W, Chan WCW: A framework for designing delivery systems. *Nat Nanotechnol.* 2020, 15:819–29. 10.1038/s41565-020-0759-5.

[11] Lan J, Zou J, Xin H, Sun J, Han T, Sun M, Niu M: Nanomedicines as disruptors or inhibitors of biofilms: Opportunities in addressing antimicrobial resistance. *Journal of Controlled Release.* 2025, 381:113589. 10.1016/j.jconrel.2025.113589.

[12] Blair JMA, Webber MA, Baylay AJ, Ogbolu DO, Piddock LJ V.: Molecular mechanisms of antibiotic resistance. *Nat Rev Microbiol.* 2015, 13:42–51. 10.1038/nrmicro3380.

[13] Wang Y, Yu Z, Ding P, et al.: Antidepressants can induce mutation and enhance persistence toward multiple antibiotics. *Proceedings of the National Academy of Sciences.* 2023, 120:. 10.1073/pnas.2208344120.

[14] Galgano M, Pellegrini F, Catalano E, et al.: Acquired Bacterial Resistance to Antibiotics and Resistance Genes: From Past to Future. *Antibiotics.* 2025, 14:222. 10.3390/antibiotics14030222.

[15] C Reygaert W: An overview of the antimicrobial resistance mechanisms of bacteria. *AIMS Microbiol.* 2018, 4:482–501. 10.3934/microbiol.2018.3.482.

[16] Blair JM, Richmond GE, Piddock LJ: Multidrug Efflux Pumps in Gram-Negative Bacteria and Their Role in Antibiotic Resistance. *Future Microbiol.* 2014, 9:1165–77. 10.2217/fmb.14.66.

[17] Deng Z, Shan Y, Pan Q, Gao X, Yan A: Anaerobic expression of the gadE-mdtEF multidrug efflux operon is primarily regulated by the two-component system ArcBA through antagonizing the H-NS mediated repression. *Front Microbiol.* 2013, 4:. 10.3389/fmicb.2013.00194.

[18] Jian Z, Zeng L, Xu T, et al.: Antibiotic resistance genes in bacteria: Occurrence, spread, and control. *J Basic Microbiol.* 2021, 61:1049–70. 10.1002/jobm.202100201.

[19] Zaki MES, Abou ElKheir N, Mofreh M: Molecular Study of Quinolone Resistance Determining Regions of gyrA Gene and parC Genes in Clinical Isolates of Acintobacter baumannii Resistant to Fluoroquinolone. *Open Microbiol* J. 2018, 12:116–22. 10.2174/1874285801812010116.

[20] Zhou M, Wang L, Wang Z, Kudinha T, Wang Y, Xu Y, Liu Z: Molecular Characterization of Penicillin-Binding Protein2x, 2b and 1a of Streptococcus pneumoniae Causing Invasive Pneumococcal Diseases in China: A Multicenter Study. *Front Microbiol.* 2022, 13:. 10.3389/fmicb.2022.838790.

[21] Tabassum F, Munir A, Shakeel F, Zia T, Zafar Khan MU, Tahir Z: Methicillin Resistance in <i>Staphylococcus aureus</i>: A Review of the mecA Gene and Its Role in Antibiotic Resistance. *Indus Journal of Bioscience Research.* 2025, 3:. 10.70749/ijbr.v3i9.2302.

[22] Wilson DN, Hauryliuk V, Atkinson GC, O'Neill AJ: Target protection as a key antibiotic resistance mechanism. *Nat Rev Microbiol.* 2020, 18:637–48. 10.1038/s41579-020-0386-z.

[23] Li W, Atkinson GC, Thakor NS, et al.: Mechanism of tetracycline resistance by ribosomal protection protein Tet(O). *Nat Commun*. 2013, 4:1477. 10.1038/ncomms2470.

[24] Svetlov MS, Syroegin EA, Aleksandrova E V., Atkinson GC, Gregory ST, Mankin AS, Polikanov YS: Structure of Erm-modified 70S ribosome reveals the mechanism of macrolide resistance. *Nat Chem Biol*. 2021, 17:412–20. 10.1038/s41589-020-00715-0.

[25] Long KS, Poehlsgaard J, Kehrenberg C, Schwarz S, Vester B: The Cfr rRNA methyltransferase confers resistance to Phenicols, Lincosamides, Oxazolidinones, Pleuromutilins, and Streptogramin A antibiotics. *Antimicrob Agents Chemother*. 2006, 50:2500–5. 10.1128/AAC.00131-06.

[26] Baran A, Kwiatkowska A, Potocki L: Antibiotics and Bacterial Resistance-A Short Story of an Endless Arms Race. *Int J Mol Sci*. 2023, 24:. 10.3390/ijms24065777.

[27] Beceiro A, Llobet E, Aranda J, et al.: Phosphoethanolamine Modification of Lipid A in Colistin-Resistant Variants of Acinetobacter baumannii Mediated by the pmrAB Two-Component Regulatory System. *Antimicrob Agents Chemother*. 2011, 55:3370–9. 10.1128/AAC.00079-11.

[28] Nguyen AH, Hood KS, Mileykovskaya E, Miller WR, Tran TT: Bacterial cell membranes and their role in daptomycin resistance: A review. *Front Mol Biosci*. 2022, 9:1035574. 10.3389/fmolb.2022.1035574.

[29] Nguyen AH, Tran TT, Panesso D, et al.: Molecular basis of cell membrane adaptation in daptomycin-resistant Enterococcus faecalis. *JCI Insight*. 2024, 9:. 10.1172/jci.insight.173836.

[30] Mora-Ochomogo M, Lohans CT: β-Lactam antibiotic targets and resistance mechanisms: from covalent inhibitors to substrates. *RSC Med Chem*. 2021, 12:1623–39. 10.1039/d1md00200g.

[31] Husna A, Rahman MdM, Badruzzaman ATM, et al.: Extended-Spectrum β-Lactamases (ESBL): Challenges and Opportunities. *Biomedicines*. 2023, 11:2937. 10.3390/biomedicines11112937.

[32] Bahr G, González LJ, Vila AJ: Metallo-β-lactamases in the Age of Multidrug Resistance: From Structure and Mechanism to Evolution, Dissemination, and Inhibitor Design. *Chem Rev*. 2021, 121:7957–8094. 10.1021/acs.chemrev.1c00138.

[33] Shaikh S, Fatima J, Shakil S, Rizvi SMohdD, Kamal MA: Antibiotic resistance and extended spectrum beta-lactamases: Types, epidemiology and treatment. *Saudi J Biol Sci*. 2015, 22:90–101. 10.1016/j.sjbs.2014.08.002.

[34] Garneau-Tsodikova S, Labby KJ: Mechanisms of resistance to aminoglycoside antibiotics: overview and perspectives. *Medchemcomm*. 2016, 7:11–27. 10.1039/C5MD00344J.

[35] Smith CA, Bhattacharya M, Toth M, Stewart NK, Vakulenko SB: Aminoglycoside resistance profile and structural architecture of the aminoglycoside acetyltransferase AAC(6')-Im. *Microb Cell*. 2017, 4:402–10. 10.15698/mic2017.12.602.

[36] Paintsil EK, Adu-Asiamah CK, Boahen KG, et al.: Genomic insights into the diversity, antimicrobial resistance and zoonotic potential of Campylobacter fetus across diverse hosts and geographies. *Microb Genom*. 2025, 11:. 10.1099/mgen.0.001446.

[37] Sharif S, Yadav AK: Bacterial biofilm and its role in antibiotic resistance. *The Microbe.* 2025, 7:100356. 10.1016/j.microb.2025.100356.

[38] Donlan RM, Costerton JW: Biofilms: survival mechanisms of clinically relevant microorganisms. *Clin Microbiol Rev.* 2002, 15:167–93. 10.1128/CMR.15.2.167-193.2002.

[39] Sahoo K, Meshram S: Biofilm Formation in Chronic Infections: A Comprehensive Review of Pathogenesis, Clinical Implications, and Novel Therapeutic Approaches. *Cureus*. Published Online First: 1 October 2024. 10.7759/cureus.70629.

[40] Kunnath AP, Suodha Suoodh M, Chellappan DK, Chellian J, Palaniveloo K: Bacterial Persister Cells and Development of Antibiotic Resistance in Chronic Infections: An Update. *Br J Biomed Sci*. 2024, 81:12958. 10.3389/bjbs.2024.12958.

[41] Grari O, Ezrari S, El Yandouzi I, et al.: A comprehensive review on biofilm-associated infections: Mechanisms, diagnostic challenges, and innovative therapeutic strategies. *The Microbe*. 2025, 8:100436. 10.1016/j.microb.2025.100436.

[42] Guillaume O, Butnarasu C, Visentin S, Reimhult E: Interplay between biofilm microenvironment and pathogenicity of Pseudomonas aeruginosa in cystic fibrosis lung chronic infection. *Biofilm*. 2022, 4:100089. 10.1016/j.bioflm.2022.100089.

[43] Abdelhamid AG, Yousef AE: Combating Bacterial Biofilms: Current and Emerging Antibiofilm Strategies for Treating Persistent Infections. *Antibiotics*. 2023, 12:1005. 10.3390/antibiotics12061005.

[44] Vanderpool EJ, Rumbaugh KP: Host-microbe interactions in chronic rhinosinusitis biofilms and models for investigation. *Biofilm*. 2023, 6:100160. 10.1016/j.bioflm.2023.100160.

[45] Abed N, Couvreur P: Nanocarriers for antibiotics: A promising solution to treat intracellular bacterial infections. *Int J Antimicrob Agents*. 2014, 43:485–96. 10.1016/j.ijantimicag.2014.02.009.

[46] Drulis-Kawa Z, Dorotkiewicz-Jach A: Liposomes as delivery systems for antibiotics. *Int J Pharm*. 2010, 387:187–98. 10.1016/j.ijpharm.2009.11.033.

[47] Huang Y, Guo X, Wu Y, Chen X, Feng L, Xie N, Shen G: Nanotechnology's frontier in combatting infectious and inflammatory diseases: prevention and treatment. *Signal Transduct Target Ther*. 2024, 9:34. 10.1038/s41392-024-01745-z.

[48] Sun X, Huang S, Ye T, et al.: Nanomaterial-powered immunotherapy: A next-generation therapeutic paradigm for bacterial infections. *Chemical Engineering Journal*. 2025, 524:169008. 10.1016/j.cej.2025.169008.

[49] Desai N, Rana D, Patel M, Bajwa N, Prasad R, Vora LK: Nanoparticle Therapeutics in Clinical Perspective: Classification, Marketed Products, and Regulatory Landscape. *Small*. 2025, 21:. 10.1002/smll.202502315.

[50] Aryal S, Jack Hu C-M, Fu V, Zhang L: Nanoparticledrug delivery enhances the cytotoxicity of hydrophobic–hydrophilic drug conjugates. *J Mater Chem*. 2012, 22:994–9. 10.1039/C1JM13834K.

[51] Aydemir D, Çakır S, Özdemir N, Ulusu NN: Evaluation of the Antimicrobial Activity of Triple Enzyme-Embedded Organic–Inorganic Hybrid Nanoflowers (hNFs) in Comparison with Powerful Antimicrobial Agent Chitosan. *Curr Microbiol*. 2024, 81:359. 10.1007/s00284-024-03884-6.

[52] Nag M, Lahiri D, Mukherjee D, et al.: Functionalized Chitosan Nanomaterials: A Jammer for Quorum Sensing. *Polymers* (Basel). 2021, 13:. 10.3390/polym13152533.

[53] Adnan SB, Maarof M, Fauzi MB, Md Fadilah NI: Antimicrobial Peptides in Wound Healing and Skin Regeneration: Dual Roles in Immunity and Microbial Defense. *Int J Mol Sci*. 2025, 26:5920. 10.3390/ijms26135920.

[54] Rajinikanth B S, Rajkumar DSR, K K, Vijayaragavan V: Chitosan-Based Biomaterial in Wound Healing: A Review. *Cureus*. 2024, 16:e55193. 10.7759/cureus.55193.

[55] Lipatov AS, Gitel'man AK, Govorkova EA, Smirnov IA: [Changes in biological and physico-chemical properties of avian influenza virus A hemagglutinin H2 during adaptation to a new host]. *Vopr Virusol*. 1995, 40:208–11.

[56] Baniasadi H: State-of-the-art in natural hydrogel-based wound dressings: Design, functionalization, and fabrication approaches. *Adv Colloid Interface Sci*. 2025, 342:103527. 10.1016/j.cis.2025.103527.

[57] Alven S, Aderibigbe BA: Chitosan-Based Scaffolds Incorporated with Silver Nanoparticles for the Treatment of Infected Wounds. *Pharmaceutics*. 2024, 16:. 10.3390/pharmaceutics16030327.

[58] Singh R, Shitiz K, Singh A: Chitin and chitosan: biopolymers for wound management. *Int Wound J*. 2017, 14:1276–89. 10.1111/iwj.12797.

[59] Naghib SM, Amiri S, Mozafari MR: Stimuli-responsive chitosan-based nanocarriers for drug delivery in wound dressing applications: A review. *Carbohydrate Polymer Technologies and Applications*. 2024, 7:100497. 10.1016/j.carpta.2024.100497.

[60] Mishra A, Aggarwal A, Khan F: Medical Device-Associated Infections Caused by Biofilm-Forming Microbial Pathogens and Controlling Strategies. *Antibiotics* (Basel). 2024, 13:. 10.3390/antibiotics13070623.

[61] Ke C-L, Deng F-S, Chuang C-Y, Lin C-H: Antimicrobial Actions and Applications of Chitosan. *Polymers* (Basel). 2021, 13:. 10.3390/polym13060904.

[62] Egorov AR, Kirichuk AA, Rubanik V V, Rubanik V V, Tskhovrebov AG, Kritchenkov AS: Chitosan and Its Derivatives: Preparation and Antibacterial Properties. *Materials* (Basel). 2023, 16:. 10.3390/ma16186076.

[63] dos Santos EMP, Martins CCB, de Oliveira Santos JV, et al.: Silver nanoparticles–chitosan composites activity against resistant bacteria: tolerance and biofilm inhibition. *Journal of Nanoparticle Research*. 2021, 23:196. 10.1007/s11051-021-05314-1.

[64] Islam MS, Elahee GMF, Fang Y, Yu X (Bill), Advincula RC, Cao C (Chase): Polylactic acid (PLA)-based multifunctional and biodegradable nanocomposites and their applications. *Compos B Eng*. 2025, 306:112842. 10.1016/j.compositesb.2025.112842.

[65] Teixeira-Santos R, Lima M, Gomes LC, Mergulhão FJ: Antimicrobial coatings based on chitosan to prevent implant-associated infections: A systematic review. *iScience*. 2021, 24:103480. 10.1016/j.isci.2021.103480.

[66] Edo GI, Mafe AN, Razooqi NF, et al.: Advances in bio-polymer coatings for probiotic microencapsulation: chitosan and beyond for enhanced stability and

controlled release. *Des Monomers Polym. 2025*, 28:1–34. 10.1080/15685551.2024.2448122.

[67] Abourehab MAS, Pramanik S, Abdelgawad MA, Abualsoud BM, Kadi A, Ansari MJ, Deepak A: Recent Advances of Chitosan Formulations in Biomedical Applications. *Int J Mol Sci*. 2022, 23:10975. 10.3390/ijms231810975.

[68] Vosoughi P, Naghib SM, Jafari T, Kangarshahi BM: Chitosan-encapsulated lipid-based nanovesicles for therapeutic applications and tissue engineering: A review. *Carbohydrate Polymer Technologies and Applications*. 2025, 10:100805. 10.1016/j.carpta.2025.100805.

[69] Khalil A, Barras A, Boukherroub R, et al.: Enhancing paracellular and transcellular permeability using nanotechnological approaches for the treatment of brain and retinal diseases. *Nanoscale Horiz*. 2024, 9:14–43. 10.1039/D3NH00306J.

[70] Herdiana Y, Wathoni N, Shamsuddin S, Muchtaridi M: Drug release study of the chitosan-based nanoparticles. *Heliyon*. 2022, 8:e08674. 10.1016/j.heliyon.2021.e08674.

[71] Tan RSL, Hassandarvish P, Chee CF, Chan LW, Wong TW: Chitosan and its derivatives as polymeric anti-viral therapeutics and potential anti-SARS-CoV-2 nanomedicine. *Carbohydr Polym*. 2022, 290:119500. 10.1016/j.carbpol.2022.119500.

[72] Gonciarz W, Balcerczak E, Brzeziński M, Jeleń A, Pietrzyk-Brzezińska AJ, Narayanan VHB, Chmiela M: Chitosan-based formulations for therapeutic applications. A recent overview. *J Biomed Sci*. 2025, 32:62. 10.1186/s12929-025-01161-7.

[73] Ma H, Zhao Y, Lu Z, et al.: Citral-loaded chitosan/carboxymethyl cellulose copolymer hydrogel microspheres with improved antimicrobial effects for plant protection. *Int J Biol Macromol*. 2020, 164:986–93. 10.1016/j.ijbiomac.2020.07.164.

[74] Jaber N, Al-Remawi M, Al-Akayleh F, Al-Muhtaseb N, Al-Adham ISI, Collier PJ: A review of the antiviral activity of Chitosan, including patented applications and its potential use against COVID-19. *J Appl Microbiol*. 2022, 132:41–58. 10.1111/jam.15202.

Chapter 5

Chitosan Under Fire: A Green Biopolymer in the Fight Against Radiation-Induced Damage

Duygu Aydemir[1]
and Nuriye Nuray Ulusu[2,3,*]
[1]Istanbul University, Institute of Child Health, Department of Pediatric Basic Sciences, Division of Medical Genetics, Istanbul, Türkiye
[2]Koç University, School of Medicine, Department of Medical Biochemistry, Sariyer, Istanbul, Türkiye
[3]Koç University Research Center for Translational Medicine (KUTTAM), Sariyer, Istanbul, Türkiye

Abstract

Chitin (β-(1→4)-poly-N-acetyl-D-glucosamine) is the most abundant aminopolysaccharide and the second most prevalent natural polymer after cellulose. It is primarily sourced from the exoskeletons of crustaceans such as crabs, lobsters, and shrimps, as well as from insects and fungal cell walls. Enzymatic or chemical deacetylation of chitin results in the conversion of chitin to chitosan. This change makes chitin a more useful biopolymer distinguished by its biocompatibility, biodegradability, film-forming ability, and non-toxic nature. These unique properties have enabled the extensive use of chitosan across a wide range of fields.

Chitosan is used in the remediation of wastewater and polluted environments due to its excellent chelating ability and adsorption capacity. It can effectively bind and remove heavy metals, dyes, and various organic pollutants from aqueous media. Moreover, it is utilized in air filtration, dust suppression, and the development of biodegradable

* Corresponding Author's Email: nulusu@ku.edu.tr

In: Chitosan
Editor: Denis Hurst
ISBN: 979-8-90134-079-0

packaging materials, aligning with global efforts toward green technologies.
Radiation exposure, particularly from ionizing sources such as gamma rays and X-rays, can induce significant biological damage, primarily by generating reactive oxygen species (ROS). Recent studies have investigated the potential of chitosan and its derivatives as radioprotective agents, which involve scavenging free radicals and promoting tissue regeneration. For instance, chitosan-based formulations have demonstrated antioxidant activity that mitigates radiation-induced cellular and DNA damage in preclinical models.
In recent years, growing environmental concerns and the increasing use of ionizing radiation across various sectors, ranging from energy production to medical diagnostics, have intensified the search for sustainable materials to lessen radiation-induced damage. Environmental science, focusing on ecological health and sustainable innovation, now intersects more frequently with radiation protection research. In this context, chitosan, a biodegradable and non-toxic biopolymer derived from natural sources such as crustacean shells, has gained considerable attention. Not only does chitosan exhibit strong chelating and antioxidant properties, but its potential to act as a biosorbent for radioactive contaminants aligns with green chemistry principles. Moreover, modified chitosan materials have shown promise in reducing oxidative stress caused by radiation exposure, thus opening avenues for their integration into eco-friendly shielding systems and post-radiation biomedical care. The convergence of radiation science and environmental sustainability makes chitosan a compelling candidate for future protective technologies.
This chapter explores the diverse applications of chitosan, with a particular focus on its roles in environmental science and radiation-related technologies, including its potential use in radioprotection and the remediation of radiation-contaminated environments.

Keywords: chitosan, radiation protection, environmental science, biopolymer, oxidative stress, radioactive waste

Introduction

Ionizing radiation (IR) has many biochemical roles in biological molecules and living organisms (Reisz et al., 2014). IR has DNA-damaging effects (Maayah et al., 2022), triggers mutations, and is recognized as a 'universal carcinogen' (Little JB, 2003), leading to the development of malformations (Lim et al., 2015). IR also affects the genetic, biochemical, physiological, and signaling systems of organisms, leading to cellular disruptions. IR exposure increases the production of reactive oxygen species (ROS), causes oxidative stress, and damage in biological systems (Gudkov et al., 2019). IR has altered the proteomic content of extracellular vesicles, DNA repair mechanisms, chromatin packaging, and protein folding (Abramowicz et al., 2019). All these harmful effects of IR at the cellular and molecular level have been shown to contribute to gene mutations, large-scale DNA deletions, interfere with DNA repair mechanisms, accelerate cell senescence, and induce cellular death mechanisms (Mavragani et al., 2019). On the other hand, understanding the radiation-induced cell death mechanisms oriented the use of IR to radiotherapy and cancer therapy. Due to the radiation dose and environmental factors, the activation of molecular pathways leads to the activation of various cell death mechanisms, including necrosis, autophagy-dependent cell death, pyroptosis, ferroptosis, and immunogenic cell death (Jiao et al., 2022). Radiotherapy is an effective method for killing tumor cells and cancer therapy; however, it often causes side effects (Mun et al., 2018). Radiation causes damage to surrounding cells, and it is toxic to normal cells. Therefore, effective compounds that prevent the adverse effects of ionizing radiation are required (Zhang et al., 2023). Some naturally occurring pharmacological compounds exhibit radioprotective effects through their radical-scavenging properties, anti-inflammatory effects, and upregulation of heat shock protein expression (Mun et al., 2018). Apigenin (Begum et al., 2012), bergenin (Dwivedi et al., 2017; Pandey et al., 2025), caffeine (Soilov et al., 1994), curcumin (Verma, 2016), delphinidin (Kim et al., 2018), selenium, vitamin E (Weiss and Landauer, 2000), propolis (Ibáñez et al., 2023), resveratrol, gallic acid, quercetin (Adnan et al., 2021), ginseng (Lee et al., 2005) are some important natural examples that protects against radiation induced damage.

Chitosan-based or chitosan derivative-based nanoparticles, drug delivery systems (Jafernik et al., 2023), gel formulations (Ourani-Pourdashti et al., 2022), nanospheres, nano capsules (Frigaard et al., 2022), hydrogels, coating systems (Masimov and Wasan, 2024), micro/nanosystems (Perinelli et al., 2018; Moghaddam et al., 2024) have key roles in health and environment (Zhang et al., 2022; Hefni, 2024). Chitosan and its derivatives have antioxidant properties and a significant role in radiotherapy. Young et al developed GdW10@CS nanospheres incorporating chitosan as a biocompatible carrier to enhance radiosensitization in hypoxic tumors. Chitosan facilitates efficient siRNA delivery and improves cellular uptake, while the system amplifies ROS generation and depletes intracellular GSH to overcome radioresistance. This approach significantly enhances the efficacy of radiotherapy by combining extrinsic and intrinsic mechanisms (Yong et al., 2017). Hydantoin-based N-halamine biocides were synthesized and incorporated into chitosan (CS) films to enhance their antibacterial and potential biomedical properties. While the addition of these compounds improved biological activity, it also reduced the thermal stability of chitosan, especially under UV-induced photooxidation. Given chitosan's biocompatibility and film-forming properties, the modified materials show promise for use in radiotherapy-related biomedical applications, where both antimicrobial function and structural performance are critical (Chylińska et al., 2019). Superparamagnetic Fe_3O_4 nanoparticles were stabilized using a chitosan-based coating to develop an effective MRI contrast agent for the diagnosis of non-alcoholic steatohepatitis (NASH). The chitosan layer not only enhanced biocompatibility and structural integrity but also preserved the magnetic properties essential for imaging. This chitosan-assisted nanoplatform highlights the material's broader potential in biomedical imaging and radiotherapy-related applications (Luo et al., 2012). Chitosan was chemically cross-linked with L-aspartic and L-glutamic acid in the presence of magnesium oxide under microwave conditions to develop advanced hemostatic biomaterials. The resulting materials demonstrated vigorous antibacterial activity, biocompatibility, and structural properties suitable for use in non-sterile wound environments. Given chitosan's regenerative and antimicrobial properties, these novel composites hold significant promise for integration into radiotherapy-related wound care and tissue repair strategies (Radwan-Pragłowska et al., 2025).

In this book chapter, the complex biological and biochemical impact of ionizing radiation is discussed, and the need for radioprotective agents that can safeguard healthy tissues during radiotherapy is addressed. Chitosan and its derivatives are natural and biodegradable biopolymers that are nontoxic and possess significant, promising material properties due to their antioxidant, antimicrobial, and regenerative properties. Due to its properties, chitosan can be applied in various biomedical applications, including a carrier material for nanodelivery systems. Recent advances highlight their role in enhancing radio sensitization, supporting wound healing, and serving as a functional carrier in imaging and drug delivery systems. By integrating chitosan into radiotherapy-related biomedical applications, it is possible to improve treatment efficacy while minimizing collateral damage. This chapter aims to explore the multifaceted potential of chitosan in combating radiation-induced damage and advancing next-generation therapeutic strategies.

Chitosan: Structure, Sources

Chitosan is a key polysaccharide derivative of chitin, obtained through deacetylation (Al-Manhel et al., 2018). Chitin (β-(1→4)-poly-N-acetyl-D-glucosamine) is the most abundant aminopolysaccharide and the second most prevalent natural polymer after cellulose (Kozma et al., 2022a). Chitosan can be obtained from various organisms, such as green mussel shells (Kaewprachu and Jaisan, 2023a), tiger shrimp (Ahing and Wid, 2016), pink shrimp shells, brown shrimp shells, crab shells, catfish shells, squid pens (Abdou et al., 2008), green mussel (Perna viridis) (Kaewprachu and Jaisan, 2023b) insects (Luo et al., 2019), mollusks (Kozma et al., 2022a), lobster (Kozma et al., 2024) blue swimmer (Portunus pelagicus) (Tissera et al., 2021). Extracting high-quality chitin from organisms with minimal waste is very crucial (Kozma et al., 2022). Acids, bases, and microorganisms are used for the extraction of chitin (Dong et al., 2023).

Chitin can be obtained from organisms, but it is not easy to deproteinize and demineralize them. These steps typically require acid treatment for chemical deproteinization and demineralization, especially for the removal of calcium carbonate (Younes and Rinaudo, 2015). Different polyol-based eutectic solvents, including ethylene glycol, glycerol, xylitol, and sorbitol, were utilized for the extraction of chitin (Zhu et al., 2024). The crucial key point in the isolation of high-quality chitin nanofibers (ChNFs) is the development of a cost-effective, environmentally friendly, and sustainable method. Huang et al used the gradient heating activated (GHA) ammonium persulfate (APS) oxidation technique to produce high-quality ChNFs (Huang et al., 2024).

Chitosan: Functional Properties

Chitosan and chitosan derivatives have biocompatible, biodegradable, and non-toxic properties. Therefore, these molecules can be used in a wide range of applications, including food, cosmetic, pharmaceutical, and biomedical industries (Kozma et al., 2022). Chitosan can be used in various biological applications, including antimicrobial, anticancer, antioxidant, antibacterial, and antifungal properties. Chitosan possesses biologically safe and environmentally friendly properties, making it highly applicable in various fields, including corrosion inhibition, catalysis, metal ion adsorption, and sensor development (Hussain and Berry, 2024), as well as a potential protective material against radiation (Azmoonfar et al., 2025a). Chitosan also serves as a promising model due to its extensive application in polymeric nanoparticle research, offering valuable insights into the role of delivery systems in stabilizing natural antioxidants and improving the efficacy of anticancer drugs (Herdiana et al., 2023). Chitosan's unique physicochemical characteristics contribute to overcoming oral drug delivery barriers, facilitating targeted delivery, minimizing side effects, and enhancing the efficiency and stability of cancer drugs in colon cancer cells (Choukaife et al., 2022) and breast cancer. Increasing the efficiency of cancer drugs is very important, for instance, breast cancer accounts for around 13% all deaths (Herdiana et al., 2023). In our previous research, we utilized the biocompatible and non-toxic properties of chitosan as a nanocarrier to develop an efficient drug delivery system for gemcitabine, while minimizing cytotoxicity in healthy cells. GEM-loaded nanoparticles not only enhanced cellular uptake but also amplified apoptotic and ferroptotic responses in pancreatic cancer cells by modulating oxidative stress and metabolic pathways (Aydemir et al., 2024b). It has been investigated that chitosan served as a benchmark antimicrobial agent, owing to its broad-spectrum activity and well-documented biocompatibility, making it widely applicable across various industries. Its comparison with enzyme-based hybrid nanoflowers highlighted chitosan's continued relevance as a potent natural polymer in the development of antimicrobial strategies targeting both Gram-positive and Gram-negative bacteria (Aydemir et al., 2024a). Chitosan has free radical scavenging properties and can protect against infrared radiation by reducing oxidative stress (Uysal Çalıcıoğlu et al., 2025).

Polysaccharide-based drug delivery systems (DDSs) are considered among the most reliable carriers for cancer therapy, owing to their favorable biological and physicochemical properties. Features such as ease of purification and functionalization, hydrophilicity, serum stability, high drug loading capacity, biocompatibility, biodegradability, enhanced bioavailability, and the ability to provide sustained and stimuli-responsive drug release make these biopolymers particularly suitable for targeted cancer treatment (Hsu et al., 2024). The unique properties of chitosan and its derivatives, including their hydrophilic nature, biocompatibility, and biodegradability, make them attractive candidates for use in the design of nanocarriers (Ghaz-Jahanian et al., 2015). Chitosan has many application areas; therefore, it must be pure and of high quality for biomedical and health applications (Aranaz et al., 2021). Chitosan has numerous biomedical applications and can serve as a key biomaterial in wound repair. Chitosan and its derivatives can be used in hydrogels for wound dressing (Zhao et al., 2023), pharmaceuticals (Pellis et al., 2022), nanoparticle-based drug delivery systems (Narmani and Jafari, 2021), intrauterine devices (Overton et al., 2024), a potential anticancer formulation in mouthwash (Maj et al., 2024), bone scaffold biomaterial (Kozusko et al., 2018), and can be used in radiation dermatitis (Haponska et al., 2021).

Chitosan and its composites are recognized as a significant biopolymer with a broad spectrum of biochemical, physiological, and biological activities, which can also combat radiation-induced damage (Das et al., 2024). In summary, recent studies have shown that chitosan and its composites make significant contributions to radiation-induced damage, wound healing, tissue engineering, drug delivery, and blood compatibility, and can be used in sutures because they possess a combination of biocompatibility, biodegradable properties, antimicrobial, anti-inflammatory, and antioxidant properties (Yadav et al., 2024). Chitosan-based nanoparticles developed from antioxidant-containing molecules, for example, curcumin, can be added to the nanoparticle. The researchers named this nanoparticle Cur-CsNPs. Cur-CsNPs were utilized in radiation-induced liver injury and evaluated for their therapeutic potential in radioprotective applications, revealing an increased therapeutic efficiency (Azmoonfar et al., 2025b).

Radiation-Induced Damage Mechanisms in Biological Tissues and Cells

Ionizing radiation-induced water splitting and production of hydroxyl radical in the presence of oxygen. The ionizing radiation increases ROS production, and ROS in turn increase cell injury. Therefore, ionizing radiation is accepted as a stress inducer (Moloudi and Haghdoost, 2025). Ionizing radiation-induced generation of ROS causes metabolic oxidative stress and cellular damage depending on radiation dose, dose-rate, and quality (Nuszkiewicz et al., 2020). Ionizing radiation also causes the generation of reactive nitrogen species (RNS), and both ROS and RNS cause ionization of molecules, modify protein structure and function, for instance, cause S-nitrosylation, and S-thiolation (Mikkelsen and Wardman, 2003). ROS activates the activity of antioxidant enzymes and antioxidant defense mechanisms, which are also linked to aging, senescence, and diseases (Sakata et al., 2015). Radiation therapy can trigger a series of effects, including cellular DNA damage responses (DDRs), cell cycle arrest, DNA repair mechanisms, and affect DDRs signalling pathways, DNA sugar, and DNA base damage (Nuszkiewicz et al., 2020), cellular pathways (Maier et al., 2016), protein synthesis (Tamaddondoust et al., 2022), regulation of protein synthesis (Braunstein et al., 2009), immunological responses (Rodríguez-Ruiz et al., 2018), mitochondria-dependent epigenetic and functional control (Averbeck and Rodriguez-Lafrasse, 2021), induce cytokine storm (Zhang et al., 2021), immune dysfunction, tissue damage, and tissue repair mechanisms (Hollingsworth et al., 2023), induce circulatory and metabolic diseases (Tapio et al., 2021), cellular senescence, aging, and age-related diseases (Wang et al., 2024). ROS, which are produced via ionizing radiation, can activate nuclear factor erythroid 2-related factor (NRF2). Consequently, NRF2 interacts with other pathways, including the Wnt/β-catenin, PPARγ, and BDNF/TrkB. These signaling pathways are related to osteogenesis, adipogenesis, and neurogenesis, respectively (Moloudi and Haghdoost, 2025). Radiation-induced lipid peroxidation (Ye et al., 2020) and mitophagy (Yang et al., 2023) trigger the cell death mechanism known as ferroptosis. Radiation-induced membrane damage also causes hemoglobin oxidation, and denaturation was measured due to the production of OH radicals (PUCHALA et al., 2004).

Understanding these radiation-induced damage mechanisms is crucial for optimizing therapeutic radiation and addressing radiation-induced problems in the design of nanoparticles, nanodrugs, and the development of therapeutic strategies (Shang et al., 2025).

Chitosan's Protective Mechanisms

Free Radical Scavenging

The amino group at the reducing end of chitosan is responsible for scavenging superoxide radicals. Fully deacetylated dimer (DD) showed the strongest DPPH (is the polysaccharide from ginseng named as 2,2-diphenyl-1-picrylhydrazyl) scavenging activity (Hao et al., 2021). Protection from free radicals is crucial for biomaterials. CM-chitosan has demonstrated a high capacity to scavenge radicals better than hyaluronic acid in the synovial fluid. This radical capacity protected synovial components from oxidative degradation. But the researchers stated that the mechanism is not clearly elucidated (Vandeweerd et al., 2021). Zou et al. developed a nanozyme-based system (S-Z@ChF) containing shikonin and zinc for the treatment of inflammatory bowel disease (IBD) caused by excessive levels of ROS and H_2S. This pH-responsive nanoplatform effectively targeted colonic inflammation sites, scavenged ROS and H_2S, suppressed the inflammatory response, and contributed to the restoration of the intestinal barrier. Additionally, compared to other nanozymes, S-Z@ChF exhibits an excellent ability to target inflamed colons and regulate free radicals (Zou et al., 2025). The pharmacological activities of chitooligosaccharides (COS) were examined, revealing that they exhibit radical scavenging activity by decreasing oxidative stress markers and inflammatory mediators, while also suppressing the activation of NF-κB (Marmouzi et al., 2019).

Chelating Metal Ions

Chitosan and chitosan derivatives can be regarded as excellent metal ion chelating agents. Therefore, chitosan and its derivatives are accepted as biocompatible, non-toxic, and environmentally friendly, with biodegradable and biocompatible properties (Verma and Quraishi, 2021; Shankar et al., 2023; Edward et al., 2024). The functionalization of chitosan improves strength, chemical stability, and antifouling properties (Edward et al., 2024). Some heavy metals are toxic and inhibit antioxidant enzymes (Ulusu et al., 2003; Tandogan and Ulusu, 2006; Tandoğan and Ulusu, 2007; Tandogan and Ulusu, 2010), damaging tissues. A chitosan-based hydrogel, named Chitosan@DOTAGA, was developed and functionalized with metal chelating properties for the treatment of complex wounds (Durand et al., 2025). Physically and chemically modified deacetylated chitosan and chitin blends and composites can remove heavy metals from water (Shankar et al., 2023). Chitosan-blended membranes can remove heavy metals from aqueous systems (Edward et al., 2024). Lin et al. developed a multifunctional polyacrylamide/chitosan hydrogel for the absorption of noxious dyes and metal ions from water, which is crucial for environmental safety (Lin et al., 2023). The removal of multiple heavy metal ions, including Cr3+, Cu2+, Mn2+, Ni2+, Pb2+, and Zn2+, using a xanthated chitosan, is an essential property for environmental protection (Yang et al., 2019).

Removal of Radioactive Waste/Solutions through the Adsorption Capacity

Chitosan nanoparticles have nano-sorbent potential and can remove toxic pollutants, radionucleotide radioactive heavy metals, and rare earth metals from wastewater (Benettayeb et al., 2023; de Araujo et al., 2023a). This molecule has a high adsorption capacity of radioactive material, for instance, removing radiocesium (Cs-137) (Mahmoud et al., 2024), (Cs-134) (Emara et al., 2023), uranium (Stopa and Yamaura, 2010), americium (de Araujo et al., 2023a), strontium (Alhokbany et al., 2022), and radiocobalt (Co-60) from aqueous media (Mahmoud et al., 2024). Chitosan-zeolite can remove caesium (Cs) from radioactive liquid (de Araujo et al., 2023b). Chitosan-coated magnetic nanoparticles can also be utilized for the rapid removal of radioactive materials from aqueous solutions (Elwakeel et al., 2014).

Immunomodulatory Effects

Humans cannot synthesize chitin or chitosan, but they can catabolize these compounds via chitin- and chitosan-degrading enzymes, such as chitinases, chitosanases, and lysozyme (Kumar and Zhang, 2019). It has been investigated that chitosan derivatives such as chitosan sulfate, carboxymethyl chitosan, and chitosan ammonium salts effectively induce the human immune system (Li et al., 2021). Chitosan stimulates the secretion of growth factors, chemokines, bioactive lipids, and inflammatory cytokines. Due to its immunostimulatory effects, it is accepted as an immunomodulatory molecule (Ghattas et al., 2025). The immunomodulatory property of chitosan is crucial because radiotherapy can suppress specific immune cells in the immune system, leading to radioresistance by altering the tumor microenvironment (Guo et al., 2023). It is widely accepted that combining radiotherapy with immunotherapy enhances the killing effect on cancer cells (Jiang et al., 2025). Chitosan-based nano-delivery systems, for instance, polymeric nanoparticles, micelles, liposomes, dendrimers, and conjugated solid lipid nanoparticles, are preferred nanosystems in cancer therapy due to their immunomodulatory, stimuli-responsive, antimicrobial, antioxidant, wound-healing, analgesic, antirheumatic, mucoadhesive, antiproliferative, and antimetastatic properties (Al-Shadidi et al., 2024).

Biomedical and Healthcare Application Areas of Chitosan

Chitosan has many application areas, but most of the applications require pure and high-quality chitosan (Aranaz et al., 2021). Chitosan and its derivatives can be used as a key biomaterial in wound repair. Chitosan and its derivatives can be incorporated into hydrogels for wound dressings (Zhao et al., 2023). The key properties of chitosan and its derivatives include biocompatibility, biodegradability, non-toxicity, anti-inflammatory, and antimicrobial properties, which collectively accelerate wound healing processes (Moeini et al., 2020). It has been investigated that chitosan-based hydrogel dressings can also regulate M2 macrophage polarization, which is crucial in inflammatory diabetic wounds to promote effective healing (Wei et al., 2024), especially in burn wound healing (Li et al., 2025). Carboxymethyl chitosan accelerates wound healing in cachectic cancer patients, reducing scar hyperplasia (Chen et al., 2025), as well as chronic wound healing (Zhang et al., 2025). Chitosan-containing injectable composite hydrogel (OCS/NX@Cur), combining natural polysaccharide and curcumin-loaded supramolecular self-assemblies, exhibited anti-inflammatory, antioxidant, and angiogenic effects and promoted tissue regeneration and skin repair (Zhao et al., 2025). These hydrogels can also be used in corneal wound healing. These chitosan-based hydrogels have the potential to promote healing, control infection, and maintain corneal transparency, offering a promising platform for next-generation ocular therapeutics (Hussain et al., 2025). Taurine has a significant effect on reducing oxidative stress (Guz et al., 2007), and taurine-loaded chitosan gel (TAU-GEL) has been applied in mice, where it was found to significantly improve wound repair by promoting collagen production and reducing free radical damage (Değim et al., 2002).

Wound Healing: Chitosan Dressing Radiation

Near-infrared (NIR) irradiation does not induce a significant temperature increase in carboxymethyl chitosan and PVP (CP), CPS, and alginate oxide. However, carboxymethyl chitosan, PVP, and OSD (CPD) hydrogel exhibit a rapid temperature increase under NIR exposure, leading to the denaturation of bacterial proteins and nucleic acids, thereby demonstrating antibacterial activity (Ouyang et al., 2024).

The effective treatment of complex wounds, specifically those caused by radiation exposure and characterized by oxidative stress and immune suppression, greatly benefits from the use of multifunctional chitosan-based biomaterials, particularly in combination with melatonin in wound healing therapies (Kulka-Kamińska et al., 2025). Wang et al., developed a keratin/chitosan (KRT/CS) film and evaluated its structural, mechanical, and therapeutic properties in a rat model of combined radiation-wound injury, showing its potential for safe and effective skin repair (Wang et al., 2025). Combined radiation and burn injuries (CRBI) can cause excessive ROS production and suppress the immune system. Shen et al., developed injectable self-healing hydrogels that are cross-linked with gallic acid-modified chitosan (CSGA)/oxidized dextran (ODex) hydrogels, with antioxidant and antibacterial properties that eliminate ROS and promote cell protection and support wound healing (Shen et al., 2023). Carboxymethyl chitosan-containing hydrogels injected locally to stop bleeding and inflammation, and exhibit antioxidant and antibacterial properties in combination with quercetin and in 808 nm near-infrared laser irradiation photothermal therapies (Xu et al., 2025). Chitosan-based biomaterial improved healing in full-thickness wounds, leading to enhanced re-epithelialization and reduced scar formation in a rat model. The presence of human hair follicle stem cells and fibroblasts in the wound site further supports the regenerative potential of these chitosan biomaterials, especially in radiation-compromised skin (Mohd Hilmi et al., 2013). Liu et al. developed biodegradable hydrogels with self-healing properties similar to those of natural tissues by combining O-carboxymethyl chitosan with other chemicals. These hydrogels not only adapt to mechanical stress and recover from damage but also support wound healing, offering promising potential for biomedical applications (Liu et al., 2022). A biocompatible hydrogel was developed by combining quaternized chitosan, hyaluronic acid, black phosphorus nanosheets, and silver-containing nanoparticles, which have a high drug-loading capacity. It was found to exhibit effective photothermal antibacterial activity under NIR irradiation and significantly augment wound healing in an acute wound model (Li et al., 2024). POLI-CHI hydrogel, derived from cashew tree (Anacardium occidentale L.) gum and chitosan, with or without low-level laser therapy (LLLT), on wound healing in rats. Results showed that chitosan-containing hydrogel, especially when combined with LLLT, enhanced wound contraction, improved scar appearance, and modulated inflammation, leading to more efficient tissue repair (Aragão-Neto et al., 2017). Chitosan films support stem cell adhesion and proliferation, facilitating sutureless tissue repair, which is crucial in regenerative medicine (Ruprai et al., 2019). An et al. developed a 4-carboxyphenylboronic acid and natural flavonoid rutin (RU)- modified chitosan containing a multifunctional hydrogel (CHI/CPBA/RU), which exhibits anti-inflammatory and antibacterial properties for skin wound healing. This CHI/CPBA/RU significantly accelerates wound healing by decreasing inflammation, increasing collagen deposition, promoting angiogenesis, and achieving 98% healing within 10 days compared to 78% in the control group (An et al., 2024). It is well established that the rotator cuff plays a crucial role in maintaining shoulder joint function (Aydemir et al., 2022). Furthermore, research has demonstrated that chitosan platelet-rich plasma implants enhance rotator cuff repair outcomes (Chevrier et al., 2022).

Recent studies have highlighted the promising role of multifunctional chitosan-based biomaterials in enhancing wound healing, particularly in challenging conditions such as radiation-induced injuries, oxidative stress, and infection (Chen et al., 2025). By integrating various components, such as herbal bioactive-loaded biopolymeric formulations or drug-like molecules, these hydrogels demonstrate improved antibacterial activity, tissue regeneration, protection against UV radiation and harmful chemicals, and reduced scarring across multiple experimental models (Jangra et al., 2025; Pintilei et al., 2025).

Conclusion

Chitosan is a naturally derived, biodegradable, nontoxic, free radical-scavenging, metal-ion-chelating, multifunctional biopolymer. Recent research on chitosan has demonstrated that this biopolymer can be utilized in various biomedical and environmental applications due to its free radical scavenging properties, including the development of nanocarriers, hydrogels, imaging agents, wound dressings, tissue engineering materials, drug delivery systems, and radiation protection and removal of radioactive waste.

When examining the properties of the chitosan molecule, several aspects must be considered. Chitosan and its analogs have numerous potential applications in various fields. Even when we attempt to combine the radioactive chitosan molecule with the chitosan molecule alone, we encounter multiple issues. The first is the use of radioactive materials in attempts to obtain different forms of the molecule, expanding its potential applications. The second is its use in cleaning up radioactive ions from environmental pollution. Another significant effect is its ability to protect cells against radiation damage, and in the management of radiation-induced injuries.

References

Abdou, E. S., Nagy, K. S. A., and Elsabee, M. Z. (2008). Extraction and Characterization of Chitin and Chitosan from Local Sources. *Bioresour Technol* 99, 1359–1367. doi: 10.1016/j.biortech.2007.01.051.

Abramowicz, A., Wojakowska, A., Marczak, L., Lysek-Gladysinska, M., Smolarz, M., Story, M. D., et al., (2019). Ionizing radiation affects the composition of the proteome of extracellular vesicles released by head-and-neck cancer cells in vitro. *J Radiat Res* 60, 289–297. doi: 10.1093/jrr/rrz001.

Adnan, M., Rasul, A., Shah, M. A., Hussain, G., Asrar, M., Riaz, A., et al., (2021). Radioprotective Role of Natural Polyphenols: From Sources to Mechanisms. *Anticancer Agents Med Chem* 22, 30–39. doi: 10.2174/18715206216662 10419095829.

Ahing, F. A., and Wid, N. (2016). Optimization of shrimp shell waste deacetylation for chitosan production. *International Journal of Advanced and Applied Sciences* 3, 31–36. doi: 10.21833/ijaas.2016.10.006.

Alhokbany, N., Ahamad, T., and Alshehri, S. M. (2022). Highly porous chitosan based magnetic polymeric nanocomposite (PNC) for the removal of radioactive, Cs(I) and Sr(II) ions from aqueous solution. *J King Saud Univ Sci* 34, 102036. doi: 10.1016/j.jksus.2022.102036.

Al Manhel, A. J., Al-Hilphy, A. R. S., and Niamah, A. K. (2018). Extraction of chitosan, characterisation and its use for water purification. *Journal of the Saudi Society of Agricultural Sciences* 17, 186–190. doi: 10.1016/j.jssas.2016.04.001.

Al-Shadidi, J., Al-Shammari, S., Al-Mutairi, D., Alkhudhair, D., Thu, H. E., and Hussain, Z. (2024). Chitosan Nanoparticles for Targeted Cancer Therapy: A Review of Stimuli-Responsive, Passive, and Active Targeting Strategies. *Int J Nanomedicine* Volume 19, 8373–8400. doi: 10.2147/IJN.S472433.

An, R., Shi, C., Tang, Y., Cui, Z., Li, Y., Chen, Z., et al., (2024). Chitosan/rutin multifunctional hydrogel with tunable adhesion, anti-inflammatory and antibacterial properties for skin wound healing. *Carbohydr Polym* 343, 122492. doi: 10.1016/j.carbpol.2024.122492.

Aragão-Neto, A. C., Soares, P. A. G., Lima-Ribeiro, M. H. M., Carvalho, E. J. A., Correia, M. T. S., and Carneiro-da-Cunha, M. G. (2017). Combined therapy using low level laser and chitosan-policaju hydrogel for wound healing. *Int J Biol Macromol* 95, 268–272. doi: 10.1016/j.ijbiomac.2016.11.019.

Aranaz, I., Alcántara, A. R., Civera, M. C., Arias, C., Elorza, B., Heras Caballero, A., et al., (2021). Chitosan: An Overview of Its Properties and Applications. *Polymers (Basel)* 13, 3256. doi: 10.3390/polym13193256.

Averbeck, D., and Rodriguez-Lafrasse, C. (2021). Role of Mitochondria in Radiation Responses: Epigenetic, Metabolic, and Signaling Impacts. *Int J Mol Sci* 22, 11047. doi: 10.3390/ijms222011047.

Aydemir, D., Çakır, S., Özdemir, N., and Ulusu, N. N. (2024a). Evaluation of the Antimicrobial Activity of Triple Enzyme-Embedded Organic–Inorganic Hybrid Nanoflowers (hNFs) in Comparison with Powerful Antimicrobial Agent Chitosan. *Curr Microbiol* 81, 359. doi: 10.1007/s00284-024-03884-6.

Aydemir, D., Eren, I., Demirhan, M., and Ulusu, N. N. (2022). Evaluation of the Cell Behavior and Growth Characteristics of the Porcine Dermal Xenograft Patch in Relation to the Surface Properties. *Front Bioeng Biotechnol* 10. doi: 10.3389/fbioe.2022.811446.

Aydemir, D., Öztürk, K., Arslan, F. B., Çalis, S., and Ulusu, N. N. (2024b). Gemcitabine-loaded chitosan nanoparticles enhanced apoptotic and ferroptotic response of gemcitabine treatment alone in the pancreatic cancer cells in vitro. *Naunyn Schmiedebergs Arch Pharmacol* 397, 9051–9066. doi: 10.1007/s00210-024-03193-6.

Azmoonfar, R., Moslehi, M., Shahbazi-Gahrouei, D., Shiri, E., and Azizi, M. (2025). Radioprotection and enhanced efficacy by curcumin-loaded chitosan nanoparticles in mitigating radiation-induced liver injury. *Biochem Biophys Res Commun* 753, 151512. doi: 10.1016/j.bbrc.2025.151512.

Begum, N., Prasad, N. R., Kanimozhi, G., and Hasan, A. Q. (2012). Apigenin ameliorates gamma radiation-induced cytogenetic alterations in cultured human blood lymphocytes. *Mutation Research/Genetic Toxicology and Environmental Mutagenesis* 747, 71–76. doi: 10.1016/j.mrgentox.2012.04.001.

Benettayeb, A., Seihoub, F. Z., Pal, P., Ghosh, S., Usman, M., Chia, C. H., et al., (2023). Chitosan Nanoparticles as Potential Nano-Sorbent for Removal of Toxic Environmental Pollutants. *Nanomaterials* 13, 447. doi: 10.3390/nano13030447.

Braunstein, S., Badura, M. L., Xi, Q., Formenti, S. C., and Schneider, R. J. (2009). Regulation of Protein Synthesis by Ionizing Radiation. *Mol Cell Biol* 29, 5645–5656. doi: 10.1128/MCB.00711-09.

Chen, G., Yang, C., Xu, X., Yang, L., Zhang, Y., Cai, C., et al., (2025). Multifunctional hydrogel dressing composed of trichosanthes polysaccharide and carboxymethyl chitosan accelerates cachectic wound healing and reduces scar hyperplasia. *Carbohydr Polym* 357, 123378. doi: 10.1016/j.carbpol.2025.123378.

Chevrier, A., Hurtig, M. B., and Lavertu, M. (2022). Chitosan-platelet-rich plasma implants improve rotator cuff repair in a large animal model: Pilot study. *J Biomater Appl* 37, 183–194. doi: 10.1177/08853282221085058.

Choukaife, H., Seyam, S., Alallam, B., Doolaanea, A. A., and Alfatama, M. (2022). Current Advances in Chitosan Nanoparticles Based Oral Drug Delivery for Colorectal Cancer Treatment. *Int J Nanomedicine* Volume 17, 3933–3966. doi: 10.2147/IJN.S375229.

Chylińska, M., Kaczmarek, H., and Burkowska-But, A. (2019). Preparation and characteristics of antibacterial chitosan films modified with N-halamine for biomedical application. *Colloids Surf B Biointerfaces* 176, 379–386. doi: 10.1016/j.colsurfb.2019.01.013.

Das, A., Ghosh, S., and Pramanik, N. (2024). Chitosan biopolymer and its composites: Processing, properties and applications- A comprehensive review. *Hybrid Advances* 6, 100265. doi: 10.1016/j.hybadv.2024.100265.

de Araujo, L. G., Medeiros, V. L., Guarnieri, G. de P., da Silva, D. A., Watanabe, T., Marumo, J. T., et al., (2023a). An adsorption agent based on chitosan–zeolite composite: environmental and radioactive liquid waste remediation. *Environmental Science: Advances* 2, 484–494. doi: 10.1039/D2VA00148A.

Değim, Z., Çelebi, N., Sayan, H., Babül, A., Erdoğan, D., and Take, G. (2002). An investigation on skin wound healing in mice with a taurine-chitosan gel formulation. *Amino Acids* 22, 187–198. doi: 10.1007/s007260200007.

Dong, Q., Qiu, W., Li, L., Tao, N., Liang Wang, A., Deng, S., et al., (2023). Extraction of chitin from white shrimp (Penaeus vannamei) shells using binary ionic liquid mixtures. *Journal of Industrial and Engineering Chemistry* 120, 529–541. doi: 10.1016/j.jiec.2023.01.004.

Durand, A., Pin, D., Rita, M., Kuznietsova, H., Dziubenko, N., Lysenko, V., et al., (2025). Chitosan@DOTAGA-based hydrogel: A metal-chelating dressing for the treatment of complex wounds. *Int J Biol Macromol* 320, 145884. doi: 10.1016/j.ijbiomac. 2025.145884.

Dwivedi, V. P., Bhattacharya, D., Yadav, V., Singh, D. K., Kumar, S., Singh, M., et al., (2017). The Phytochemical Bergenin Enhances T Helper 1 Responses and Anti-Mycobacterial Immunity by Activating the MAP Kinase Pathway in Macrophages. *Front Cell Infect Microbiol* 7. doi: 10.3389/fcimb.2017.00149.

Edward, K., Yuvaraj, K. M., and Kapoor, A. (2024). Chitosan-blended membranes for heavy metal removal from aqueous systems: A review of synthesis, separation mechanism, and performance. *Int J Biol Macromol* 279, 134996. doi: 10.1016/j. ijbiomac.2024.134996.

Elwakeel, K. Z., Atia, A. A., and Guibal, E. (2014). Fast removal of uranium from aqueous solutions using tetraethylenepentamine modified magnetic chitosan resin. *Bioresour Technol* 160, 107–114. doi: 10.1016/j.biortech.2014.01.037.

Emara, A. M., Elsharma, E. M., and Abdelmonem, I. M. (2023). Adsorption of radioactive cesium using synthesized chitosan-g-poly(acrylic acid/N-vinylcaprolactam) by γ-irradiation. *Radiation Physics and Chemistry* 208, 110892. doi: 10.1016/j.radphy schem.2023.110892.

Frigaard, J., Jensen, J. L., Galtung, H. K., and Hiorth, M. (2022). The Potential of Chitosan in Nanomedicine: An Overview of the Cytotoxicity of Chitosan Based Nanoparticles. *Front Pharmacol* 13. doi: 10.3389/fphar.2022.880377.

Ghattas, M., Dwivedi, G., Chevrier, A., Horn-Bourque, D., Alameh, M.-G., and Lavertu, M. (2025). Chitosan immunomodulation: insights into mechanisms of action on immune cells and signaling pathways. *RSC Adv* 15, 896–909. doi: 10.1039/D4R A08406C.

Ghaz-Jahanian, M. A., Abbaspour-Aghdam, F., Anarjan, N., Berenjian, A., and Jafarizadeh-Malmiri, H. (2015). Application of Chitosan-Based Nanocarriers in

Tumor-Targeted Drug Delivery. *Mol Biotechnol* 57, 201–218. doi: 10.1007/s12033-014-9816-3.

Gudkov, S. V., Grinberg, M. A., Sukhov, V., and Vodeneev, V. (2019). Effect of ionizing radiation on physiological and molecular processes in plants. *J Environ Radioact* 202, 8–24. doi: 10.1016/j.jenvrad.2019.02.001.

Guo, S., Yao, Y., Tang, Y., Xin, Z., Wu, D., Ni, C., et al., (2023). Radiation-induced tumor immune microenvironments and potential targets for combination therapy. *Signal Transduct Target Ther* 8, 205. doi: 10.1038/s41392-023-01462-z.

Guz, G., Oz, E., Lortlar, N., Ulusu, N. N., Nurlu, N., Demirogullari, B., et al., (2007). The effect of taurine on renal ischemia/reperfusion injury. *Amino Acids* 32, 405–411. doi: 10.1007/s00726-006-0383-1.

Hao, W., Li, K., Ma, Y., Li, R., Xing, R., Yu, H., et al., (2021). Preparation and Antioxidant Activity of Chitosan Dimers with Different Sequences. *Mar Drugs* 19, 366. doi: 10.3390/md19070366.

Haponska, M., Modrakowska, P., Balik, K., Bajek, A., Coloigner, A., Tylkowski, B., et al., (2021). Medical Plaster Enhancement by Coating with Cistus L. Extracts within a Chitosan Matrix: From Natural Complexity to Health Care Simplicity. *Materials* 14, 582. doi: 10.3390/ma14030582.

Hefni, H. H. H. (2024). Functional chitosan surfactants for corrosion inhibition in petroleum industry. *Int J Biol Macromol* 278, 134704. doi: 10.1016/j.ijbiomac.2024.134704.

Herdiana, Y., Husni, P., Nurhasanah, S., Shamsuddin, S., and Wathoni, N. (2023). Chitosan-Based Nano Systems for Natural Antioxidants in Breast Cancer Therapy. *Polymers (Basel)* 15, 2953. doi: 10.3390/polym15132953.

Hollingsworth, B. A., Aldrich, J. T., Case, C. M., DiCarlo, A. L., Hoffman, C. M., Jakubowski, A. A., et al., (2023). Immune Dysfunction from Radiation Exposure. *Radiat Res* 200. doi: 10.1667/RADE-22-00004.1.

Hsu, C.-Y., Allela, O. Q. B., Hussein, A. M., Mustafa, M. A., Kaur, M., Alaraj, M., et al., (2024). Recent advances in polysaccharide-based drug delivery systems for cancer therapy: a comprehensive review. *Artif Cells Nanomed Biotechnol* 52, 561–586. doi: 10.1080/21691401.2024.2436350.

Huang, S., Liu, T., Liu, Y., Duan, Y., and Zhang, J. (2024). Gradient heating activated ammonium persulfate oxidation for efficient preparation of high-quality chitin nanofibers. *Carbohydr Polym* 340, 122308. doi: 10.1016/j.carbpol.2024.122308.

Hussain, M. S., Alfawaz Altamimi, A. S., Afzal, M., Babu, M. A., Goyal, K., R, R., et al., (2025). Chitosan-based hydrogels: Revolutionizing corneal wound healing with antibacterial and regenerative properties. *Exp Eye Res* 258, 110503. doi: 10.1016/j.exer.2025.110503.

Hussain, S., and Berry, S. (2024). A review study on green synthesis of chitosan derived schiff bases and their applications. *Carbohydr Res* 535, 109002. doi: 10.1016/j.carres.2023.109002.

Ibáñez, B., Melero, A., Montoro, A., San Onofre, N., and Soriano, J. M. (2023). Radioprotective Effects from Propolis: A Review. *Molecules* 28, 5842. doi: 10.3390/molecules28155842.

Jafernik, K., Ładniak, A., Blicharska, E., Czarnek, K., Ekiert, H., Wiącek, A. E., et al., (2023). Chitosan-Based Nanoparticles as Effective Drug Delivery Systems—A review. *Molecules* 28, 1963. doi: 10.3390/molecules28041963.

Jangra, N., Singla, A., Puri, V., Dheer, D., Chopra, H., Malik, T., et al., (2025). Herbal bioactive-loaded biopolymeric formulations for wound healing applications. *RSC Adv* 15, 12402–12442. doi: 10.1039/D4RA08604J.

Jiang, J., Li, H., Ma, Q., Liu, J., Ren, F., Song, Y., et al., (2025). Synergies between radiotherapy and immunotherapy: a systematic review from mechanism to clinical application. *Front Immunol* 16. doi: 10.3389/fimmu.2025.1554499.

Jiao, Y., Cao, F., and Liu, H. (2022). Radiation-induced Cell Death and Its Mechanisms. *Health Phys* 123, 376–386. doi: 10.1097/HP.0000000000001601.

Kaewprachu, P., and Jaisan, C. (2023a). Physicochemical Properties of Chitosan from Green Mussel Shells (Perna viridis): A Comparative Study. *Polymers (Basel)* 15, 2816. doi: 10.3390/polym15132816.

Kim, H. M., Kim, S. H., and Kang, B. S. (2018). Radioprotective effects of delphinidin on normal human lung cells against proton beam exposure. *Nutr Res Pract* 12, 41. doi: 10.4162/nrp.2018.12.1.41.

Kozma, M., Acharya, B., and Bissessur, R. (2022a). Chitin, Chitosan, and Nanochitin: Extraction, Synthesis, and Applications. *Polymers (Basel)* 14, 3989. doi: 10.3390/polym14193989.

Kozma, M., Acharya, B., and Bissessur, R. (2024). Chemical extraction of chitin from American lobster (Homarus americanus) shells optimized through response surface methodology. *Int J Biol Macromol* 256, 128462. doi: 10.1016/j.ijbiomac.2023.128462.

Kozusko, S. D., Riccio, C., Goulart, M., Bumgardner, J., Jing, X. L., and Konofaos, P. (2018). Chitosan as a Bone Scaffold Biomaterial. *Journal of Craniofacial Surgery* 29, 1788–1793. doi: 10.1097/SCS.0000000000004909.

Kulka-Kamińska, K., Brudzyńska, P., Okura, M., Ishii, T., Skala, M., Reiter, R. J., et al., (2025). Melatonin/Chitosan Biomaterials for Wound Healing and Beyond: A Multifunctional Therapeutic Approach. *Int J Mol Sci* 26, 5918. doi: 10.3390/ijms26135918.

Kumar, A., and Zhang, K. Y. J. (2019). "Human Chitinases: Structure, Function, and Inhibitor Discovery," 221–251. doi: 10.1007/978-981-13-7318-3_11.

Lee, T.-K., Johnke, R. M., Allison, R. R., O'Brien, K. F., and Dobbs, L. J. (2005). Radioprotective potential of ginseng. *Mutagenesis* 20, 237–243. doi: 10.1093/mutage/gei041.

Li, S., Pan, W., Zhang, M., Song, K., Zhou, Z., Zhao, Q., et al., (2025). Chitosan-Based Dressing Materials for Burn Wound Healing. *Polymers (Basel)* 17, 1647. doi: 10.3390/polym17121647.

Li, T., Zhao, J., Yue, Y., Han, B., Wang, W., Zhang, H., et al., (2024). Preparation of nano-silver containing black phosphorus based on quaternized chitosan hydrogel and evaluating its effect on skin wound healing. *Int J Biol Macromol* 268, 131950. doi: 10.1016/j.ijbiomac.2024.131950.

Li, X., Xing, R., Xu, C., Liu, S., Qin, Y., Li, K., et al., (2021). Immunostimulatory effect of chitosan and quaternary chitosan: A review of potential vaccine adjuvants. *Carbohydr Polym* 264, 118050. doi: 10.1016/j.carbpol.2021.118050.

Lim, H., Agopian, A. J., Whitehead, L. W., Beasley, C. W., Langlois, P. H., Emery, R. J., et al., (2015). Maternal occupational exposure to ionizing radiation and major structural birth defects. *Birth Defects Res A Clin Mol Teratol* 103, 243–254. doi: 10.1002/bdra.23340.

Lin, X., Liu, Z., Chen, R., Hou, Y., Lu, R., Li, S., et al., (2023). A multifunctional polyacrylamide/chitosan hydrogel for dyes adsorption and metal ions detection in water. *Int J Biol Macromol* 246, 125613. doi: 10.1016/j.ijbiomac.2023.125613.

Little JB (2003). Experimental Radiation-Induced Carcinogenesis. *In: Kufe DW, Pollock RE, Weichselbaum RR, et al., editors. Holland-Frei Cancer Medicine. 6th edition. Hamilton (ON): BC Decker.*

Liu, Y., Lin, S.-H., Chuang, W.-T., Dai, N.-T., and Hsu, S. (2022). Biomimetic Strain-Stiffening in Chitosan Self-Healing Hydrogels. *ACS Appl Mater Interfaces* 14, 16032–16046. doi: 10.1021/acsami.2c01720.

Luo, Q., Wang, Y., Han, Q., Ji, L., Zhang, H., Fei, Z., et al., (2019). Comparison of the physicochemical, rheological, and morphologic properties of chitosan from four insects. *Carbohydr Polym* 209, 266–275. doi: 10.1016/j.carbpol.2019.01.030.

Luo, X., Song, X., Zhu, A., Si, Y., Ji, L., Ma, Z., et al., (2012). Development of poly(vinyl acetate-methylacrylic acid)/chitosan/Fe3O4 nanoparticles for the diagnosis of non-alcoholic steatohepatitis with magnetic resonance imaging. *J Mater Sci Mater Med* 23, 3075–3082. doi: 10.1007/s10856-012-4763-7.

Maayah, Y., Nusrat, H., Pang, G., and Tambasco, M. (2022). Assessing the DNA Damaging Effectiveness of Ionizing Radiation Using Plasmid DNA. *Int J Mol Sci* 23, 12459. doi: 10.3390/ijms232012459.

Mahmoud, H. H., Eskander, S. B., and Saleh, H. M. (2024). Biosorption Capability of Chitosan for Removal of Cs-137 and/or Co-60 from Radioactive Waste Solution Simulates. *Sustainability* 16, 1104. doi: 10.3390/su16031104.

Maier, P., Hartmann, L., Wenz, F., and Herskind, C. (2016). Cellular Pathways in Response to Ionizing Radiation and Their Targetability for Tumor Radiosensitization. *Int J Mol Sci* 17, 102. doi: 10.3390/ijms17010102.

Maj, M., Tylkowski, B., Konopka, P., Woźniak-Budych, M., Staszak, K., Staszak, M., et al., (2024). Advancing oral health: Harnessing the potential of chitosan and polyphenols in innovative mouthwash formulation. *Biomedicine & Pharmacotherapy* 175, 116654. doi: 10.1016/j.biopha.2024.116654.

Marmouzi, I., Ezzat, S. M., Salama, M. M., Merghany, R. M., Attar, A. M., EL-Desoky, A. M., et al., (2019). Recent Updates in Pharmacological Properties of Chitooligosaccharides. *Biomed Res Int* 2019, 1–16. doi: 10.1155/2019/4568039.

Masimov, R., and Wasan, E. K. (2024). Chitosan non-particulate vaccine delivery systems. *Journal of Pharmacy & Pharmaceutical Sciences* 27. doi: 10.3389/jpps.2024.12921.

Mavragani, I. V., Nikitaki, Z., Kalospyros, S. A., and Georgakilas, A. G. (2019). Ionizing Radiation and Complex DNA Damage: From Prediction to Detection Challenges and Biological Significance. *Cancers (Basel)* 11, 1789. doi: 10.3390/cancers11111789.

Mikkelsen, R. B., and Wardman, P. (2003). Biological chemistry of reactive oxygen and nitrogen and radiation-induced signal transduction mechanisms. *Oncogene* 22, 5734–5754. doi: 10.1038/sj.onc.1206663.

Moeini, A., Pedram, P., Makvandi, P., Malinconico, M., and Gomez d'Ayala, G. (2020). Wound healing and antimicrobial effect of active secondary metabolites in chitosan-based wound dressings: A review. *Carbohydr Polym* 233, 115839. doi: 10.1016/j.carbpol.2020.115839.

Moghaddam, F. D., Zare, E. N., Hassanpour, M., Bertani, F. R., Serajian, A., Ziaei, S. F., et al., (2024). Chitosan-based nanosystems for cancer diagnosis and therapy: Stimuli-responsive, immune response, and clinical studies. *Carbohydr Polym* 330, 121839. doi: 10.1016/j.carbpol.2024.121839.

Mohd Hilmi, A. B., Halim, A. S., Jaafar, H., Asiah, A. B., and Hassan, A. (2013). Chitosan Dermal Substitute and Chitosan Skin Substitute Contribute to Accelerated Full-Thickness Wound Healing in Irradiated Rats. *Biomed Res Int* 2013, 1–13. doi: 10.1155/2013/795458.

Moloudi, K., and Haghdoost, S. (2025). An Update on Role of Ionizing Radiation to Enhance Proliferation and Differentiation of Normal Stem Cells via Activation of NRF2 Pathway: Review. *Antioxidants* 14, 986. doi: 10.3390/antiox14080986.

Mun, G. I., Kim, S., Choi, E., Kim, C. S., and Lee, Y.-S. (2018). Pharmacology of natural radioprotectors. *Arch Pharm Res* 41, 1033–1050. doi: 10.1007/s12272-018-1083-6.

Narmani, A., and Jafari, S. M. (2021). Chitosan-based nanodelivery systems for cancer therapy: Recent advances. *Carbohydr Polym* 272, 118464. doi: 10.1016/j.carbpol.2021.118464.

Nuszkiewicz, J., Woźniak, A., and Szewczyk-Golec, K. (2020). Ionizing Radiation as a Source of Oxidative Stress—The Protective Role of Melatonin and Vitamin D. *Int J Mol Sci* 21, 5804. doi: 10.3390/ijms21165804.

Ourani-Pourdashti, S., Mirzaei, E., Heidari, R., Ashrafi, H., and Azadi, A. (2022). Preparation and evaluation of niosomal chitosan-based in situ gel formulation for direct nose-to-brain methotrexate delivery. *Int J Biol Macromol* 213, 1115–1126. doi: 10.1016/j.ijbiomac.2022.06.031.

Ouyang, Y., Su, X., Zheng, X., Zhang, L., Chen, Z., Yan, Q., et al., (2024). Mussel-inspired "all-in-one" sodium alginate/carboxymethyl chitosan hydrogel patch promotes healing of infected wound. *Int J Biol Macromol* 261, 129828. doi: 10.1016/j.ijbiomac.2024.129828.

Overton, E., D'Alton, M., and Goffman, D. (2024). Intrauterine devices in the management of postpartum hemorrhage. *Am J Obstet Gynecol* 230, S1076–S1088. doi: 10.1016/j.ajog.2023.08.015.

Pandey, P., Lakhanpal, S., Mahmood, D., Kang, H. N., Kim, B., Kang, S., et al., (2025). Bergenin, a bioactive flavonoid: advancements in the prospects of anticancer mechanism, pharmacokinetics and nanoformulations. *Front Pharmacol* 15. doi: 10.3389/fphar.2024.1481587.

Pellis, A., Guebitz, G. M., and Nyanhongo, G. S. (2022). Chitosan: Sources, Processing and Modification Techniques. *Gels* 8, 393. doi: 10.3390/gels8070393.

Perinelli, D. R., Fagioli, L., Campana, R., Lam, J. K. W., Baffone, W., Palmieri, G. F., et al., (2018). Chitosan-based nanosystems and their exploited antimicrobial activity.

European Journal of Pharmaceutical Sciences 117, 8–20. doi: 10.1016/j.ejps.2018.01.046.

Pintilei, P. S., Binaymotlagh, R., Chronopoulou, L., and Palocci, C. (2025). The Role of Natural Hydrogels in Enhancing Wound Healing: From Biomaterials to Bioactive Therapies. *Pharmaceutics* 17, 1243. doi: 10.3390/pharmaceutics17101243.

PUCHALA, M., SZWEDA-LEWANDOWSKA, Z., and KIEFER, J. (2004). The Influence of Radiation Quality on Radiation-induced Hemolysis and Hemoglobin Oxidation of Human Erythrocytes. *J Radiat Res* 45, 275–279. doi: 10.1269/jrr.45.275.

Radwan-Pragłowska, J., Bąk, P., Janus, Ł., Sierakowska-Byczek, A., Radomski, P., Kramek, A., et al., (2025). Synthesis and Characterization of Magnesium Oxide-Enhanced Chitosan-Based Hemostatic Gels with Antibacterial Properties: Role of Amino Acids and Crosslinking. *Molecules* 30, 1496. doi: 10.3390/molecules30071496.

Reisz, J. A., Bansal, N., Qian, J., Zhao, W., and Furdui, C. M. (2014). Effects of Ionizing Radiation on Biological Molecules—Mechanisms of Damage and Emerging Methods of Detection. *Antioxid Redox Signal* 21, 260–292. doi: 10.1089/ars.2013.5489.

Rodríguez-Ruiz, M. E., Vanpouille-Box, C., Melero, I., Formenti, S. C., and Demaria, S. (2018). Immunological Mechanisms Responsible for Radiation-Induced Abscopal Effect. *Trends Immunol* 39, 644–655. doi: 10.1016/j.it.2018.06.001.

Ruprai, H., Romanazzo, S., Ireland, J., Kilian, K., Mawad, D., George, L., et al., (2019). Porous Chitosan Films Support Stem Cells and Facilitate Sutureless Tissue Repair. *ACS Appl Mater Interfaces* 11, 32613–32622. doi: 10.1021/acsami.9b09123.

Sakata, K., Kondo, T., Mizuno, N., Shoji, M., Yasui, H., Yamamori, T., et al., (2015). Roles of ROS and PKC-βII in ionizing radiation-induced eNOS activation in human vascular endothelial cells. *Vascul Pharmacol* 70, 55–65. doi: 10.1016/j.vph.2015.03.016.

Shang, Y., Hu, X., Ren, M., Ma, L., Zhao, X., Gao, C., et al., (2025). Understanding the toxicity induced by radiation-triggered neuroinflammation and the on-demand design of targeted peptide nanodrugs. *Signal Transduct Target Ther* 10, 286. doi: 10.1038/s41392-025-02375-9.

Shankar, S., Joshi, S., and Srivastava, R. K. (2023). A review on heavy metal biosorption utilizing modified chitosan. *Environ Monit Assess* 195, 1350. doi: 10.1007/s10661-023-11963-7.

Shen, J., Jiao, W., Chen, Z., Wang, C., Song, X., Ma, L., et al., (2023). Injectable multifunctional chitosan/dextran-based hydrogel accelerates wound healing in combined radiation and burn injury. *Carbohydr Polym* 316, 121024. doi: 10.1016/j.carbpol.2023.121024.

Soilov, L. M., Mullenders, L. H. F., and Natarajan, A. T. (1994). Caffeine potentiates or protects against radiation-induced DNA and chromosal damage in human lymphocytes depending on temperature and concentration. *Mutation Research/Fundamental and Molecular Mechanisms of Mutagenesis* 311, 169–174. doi: 10.1016/0027-5107(94)90174-0.

Stopa, L. C. B., and Yamaura, M. (2010). Uranium removal by chitosan impregnated with magnetite nanoparticles: adsorption and desorption. *International Journal of Nuclear Energy Science and Technology* 5, 283. doi: 10.1504/IJNEST.2010.035538.

Tamaddondoust, R. N., Wang, Y., Jafarnejad, S. M., Graber, T. E., and Alain, T. (2022). The highs and lows of ionizing radiation and its effects on protein synthesis. *Cell Signal* 89, 110169. doi: 10.1016/j.cellsig.2021.110169.

Tandogan, B., and Ulusu, N. N. (2006). Effects of cadmium and zinc ions on purified lamb kidney cortex glucose-6-phosphate dehydrogenase activity. *J Enzyme Inhib Med Chem* 21, 225–230. doi: 10.1080/14756360500480533.

Tandoğan, B., and Ulusu, N. N. (2007). The inhibition kinetics of yeast glutathione reductase by some metal ions. *J Enzyme Inhib Med Chem* 22, 489–495. doi: 10.1080/14756360601162147.

Tandogan, B., and Ulusu, N. N. (2010). Comparative *In Vitro* Effects of Some Metal Ions On Bovine Kidney Cortex Glutathione Reductase. *Prep Biochem Biotechnol* 40, 405–411. doi: 10.1080/10826068.2010.525400.

Tapio, S., Little, M. P., Kaiser, J. C., Impens, N., Hamada, N., Georgakilas, A. G., et al., (2021). Ionizing radiation-induced circulatory and metabolic diseases. *Environ Int* 146, 106235. doi: 10.1016/j.envint.2020.106235.

Tissera, W. M. J. C., Rathnayake, S. I., Abeyrathne, E. D. N. S., and Nam, K.-C. (2021). An improved extraction and purification method for obtaining high-quality chitin and chitosan from blue swimmer (Portunus pelagicus) crab shell waste. *Food Sci Biotechnol* 30, 1645–1655. doi: 10.1007/s10068-021-01002-x.

Ulusu, N. N., Acan, N. L., Turan, B., and Tezcan, E. F. (2003). Inhibition of Glutathione Reductase by Cadmium Ion in Some Rabbit Tissues and the Protective Role of Dietary Selenium. *Biol Trace Elem Res* 91, 151–156. doi: 10.1385/BTER:91:2:151.

Uysal Çalıcıoğlu, P., Akpınar, F., Gençdağ Şensoy, K., Çalışkan, Ş. G., and Muti, M. (2025). Determination of the damaging effect of IR radiation on DNA by electrochemical methods and investigation of this effect in the presence of chitosan. *Analytical Sciences* 41, 783–792. doi: 10.1007/s44211-025-00727-7.

Vandeweerd, J.-M., Innocenti, B., Rocasalbas, G., Gautier, S. E., Douette, P., Hermitte, L., et al., (2021). Non-clinical assessment of lubrication and free radical scavenging of an innovative non-animal carboxymethyl chitosan biomaterial for viscosupplementation: An *in-vitro* and *ex-vivo* study. *PLoS One* 16, e0256770. doi: 10.1371/journal.pone.0256770.

Verma, C., and Quraishi, M. A. (2021). Chelation capability of chitosan and chitosan derivatives: Recent developments in sustainable corrosion inhibition and metal decontamination applications. *Current Research in Green and Sustainable Chemistry* 4, 100184. doi: 10.1016/j.crgsc.2021.100184.

Verma, V. (2016). Relationship and interactions of curcumin with radiation therapy. *World J Clin Oncol* 7, 275. doi: 10.5306/wjco.v7.i3.275.

Wang, H., Tong, J., and Cao, Y. (2024). Non-ionizing radiation-induced cellular senescence and age-related diseases. *Radiat Med Prot* 5, 68–74. doi: 10.1016/j.radmp.2024.04.002.

Wang, Y., Xin, T., Deng, H., Chen, J., Tang, S., Liu, L., et al., (2025). Keratin/chitosan film promotes wound healing in rats with combined radiation-wound injury. *J Mater Sci Mater Med* 36, 15. doi: 10.1007/s10856-025-06860-z.

Wei, X., Liu, C., Li, Z., Gu, Z., Yang, J., and Luo, K. (2024). Chitosan-based hydrogel dressings for diabetic wound healing via promoting M2 macrophage-polarization. *Carbohydr Polym* 331, 121873. doi: 10.1016/j.carbpol.2024.121873.

WEISS, J. F., and LANDAUER, M. R. (2000). Radioprotection by Antioxidants [a]. *Ann N Y Acad Sci* 899, 44–60. doi: 10.1111/j.1749-6632.2000.tb06175.x.

Xu, Q., Su, W., Huang, C., Zhong, H., Huo, L., Cai, J., et al., (2025). Multifunctional Polysaccharide Self-Healing Wound Dressing: NIR-Responsive Carboxymethyl Chitosan / Quercetin Hydrogel. *Adv Healthc Mater* 14. doi: 10.1002/adhm.20240 3267.

Yadav, H., Malviya, R., and Kaushik, N. (2024). Chitosan in biomedicine: A comprehensive review of recent developments. *Carbohydrate Polymer Technologies and Applications* 8, 100551. doi: 10.1016/j.carpta.2024.100551.

Yang, K., Wang, G., Liu, F., Wang, X., and Chen, X. (2019). Removal of multiple heavy metal ions using a macromolecule chelating flocculant xanthated chitosan. *Water Science and Technology* 79, 2289–2297. doi: 10.2166/wst.2019.230.

Yang, P., Li, J., Zhang, T., Ren, Y., Zhang, Q., Liu, R., et al., (2023). Ionizing radiation-induced mitophagy promotes ferroptosis by increasing intracellular free fatty acids. *Cell Death Differ* 30, 2432–2445. doi: 10.1038/s41418-023-01230-0.

Ye, L. F., Chaudhary, K. R., Zandkarimi, F., Harken, A. D., Kinslow, C. J., Upadhyayula, P. S., et al., (2020). Radiation-Induced Lipid Peroxidation Triggers Ferroptosis and Synergizes with Ferroptosis Inducers. *ACS Chem Biol* 15, 469–484. doi: 10.1021/acschembio.9b00939.

Yong, Y., Zhang, C., Gu, Z., Du, J., Guo, Z., Dong, X., et al., (2017). Polyoxometalate-Based Radiosensitization Platform for Treating Hypoxic Tumors by Attenuating Radioresistance and Enhancing Radiation Response. *ACS Nano* 11, 7164–7176. doi: 10.1021/acsnano.7b03037.

Younes, I., and Rinaudo, M. (2015). Chitin and Chitosan Preparation from Marine Sources. Structure, Properties and Applications. *Mar Drugs* 13, 1133–1174. doi: 10.3390/md 13031133.

Zhang, C., Liang, Z., Ma, S., and Liu, X. (2021). Radiotherapy and Cytokine Storm: Risk and Mechanism. *Front Oncol* 11. doi: 10.3389/fonc.2021.670464.

Zhang, H., Gao, X., Dai, W. Q., Cui, N., Liu, J. Q., Wang, M. G., et al., (2025). Lutein-loaded multifunctional hydrogel dressing based on carboxymethyl chitosan for chronic wound healing. *Int J Biol Macromol* 300, 140219. doi: 10.1016/j.ijbiomac. 2025.140219.

Zhang, R.-X., Wu, Z.-W., Zhang, S.-J., Wei, H.-M., Hua, C.-W., Li, L., et al., (2022). Gene cloning and molecular characterization of a thermostable chitosanase from Bacillus cereus TY24. *BMC Biotechnol* 22, 30. doi: 10.1186/s12896-022-00762-6.

Zhang, Y., Huang, Y., Li, Z., Wu, H., Zou, B., and Xu, Y. (2023). Exploring Natural Products as Radioprotective Agents for Cancer Therapy: Mechanisms, Challenges, and Opportunities. *Cancers (Basel)* 15, 3585. doi: 10.3390/cancers15143585.

Zhao, J., Qiu, P., Wang, Y., Wang, Y., Zhou, J., Zhang, B., et al., (2023). Chitosan-based hydrogel wound dressing: From mechanism to applications, a review. *Int J Biol Macromol* 244, 125250. doi: 10.1016/j.ijbiomac.2023.125250.

Zhao, M., Xiang, J., Meng, Y., Sun, H., Yang, W., Li, Z., et al., (2025). Astragalus Polysaccharide Hydrogels with Drug-Carrying Super Self-Assembly from Natural Herbs Promote Wound Healing. *ACS Nano* 19, 21571–21588. doi: 10.1021/acsnano.5c03744.

Zhu, H., Chen, S., Yang, T., Hu, X., Cai, W., Wang, X., et al., (2024). Comparative efficiency of different polyol-based deep eutectic solvents for chitin extraction from lobster shells. *Int J Biol Macromol* 277, 134425. doi: 10.1016/j.ijbiomac.2024.134425.

Zou, J., Ding, W., Chen, Y., Pan, X., Fu, X., Zheng, S., et al., (2025). Incorporating zinc coordination driven nanozyme into chitosan and hyaluronic acid based nanoplatform for scavenging H2S/ROS in managing inflammatory bowel disease. *Carbohydr Polym* 361, 123501. doi: 10.1016/j.carbpol.2025.123501.

Chapter 6

The Versatility of Chitosan in the Medical and Environmental Fields: From Drug Delivery to Wastewater Treatment

Iare S. Ribeiro[1]
Marcos V. S. Pereira[1]
Wesley C. P. Aquino[1]
and Jemmyson R. de Jesus[1,2,*]
[1]Department of Chemistry, Federal University of Viçosa, Viçosa, Minas Gerais, Brazil
[2]National Institute of Science and Technology of Bioanalytics-Lauro Kubota (INCTBio-LK), Campinas, São Paulo, Brazil

Abstract

Chitosan, a natural biopolymer obtained by the deacetylation of chitin, has emerged as a highly valuable material due to its exceptional characteristics, including biocompatibility, biodegradability, film-forming ability, antimicrobial activity, and low toxicity. These properties allow for its wide application in various scientific and technological fields. In medicine, its versatility makes it a promising candidate for applications such as controlled release of drugs and active compounds, wound healing, and tissue engineering, where it can be used in the form of hydrogels, nanoparticles, and scaffolds. In the environmental field, chitosan is used in water purification, heavy metal adsorption, and pollutant removal, contributing to the mitigation of ecological challenges. In these contexts, this chapter provides a comprehensive

* Corresponding Author's Email: jemmyson.jesus@ufv.br

In: Chitosan
Editor: Denis Hurst
ISBN: 979-8-90134-079-0

overview of the diverse applications of chitosan in the medical and environmental fields, highlighting challenges and prospects in the field.

Keywords: biotechnology, drug delivery system, emerging contaminants, environmental monitoring

Introduction

Chitosan (CS, Figure 1) is a high-value biopolymer derived from the partial deacetylation of chitin, a naturally abundant polysaccharide found in the exoskeletons of crustaceans and insects, and in the cell walls of fungi and algae. Its chemical modification results in solubility in slightly acidic solutions, facilitating its processing and application in various fields.

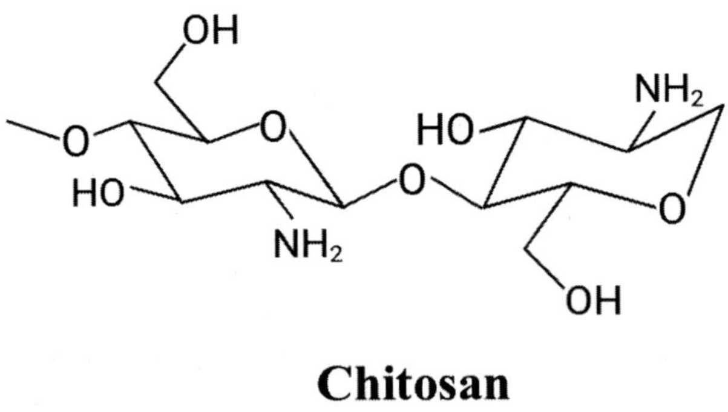

Figure 1. Chemical structure of chitosan.

The history of CS dates to the 19th century. Chitin was first identified in 1811 by Henri Braconnot and later named by Odier in 1823 (Khoushab & Yamabhai, 2010). In 1859, C. Rouget converted chitin into a soluble form, what is now known as chitosan, by treating it with concentrated alkali. Although initially poorly understood, this discovery laid the foundation for several applications in different fields. In the late 20th century, advances in processing and characterization methods allowed CS to emerge as a functional biomaterial, with growing interest in biomedical applications (Khoushab & Yamabhai, 2010). CS is currently recognized for its biodegradability, biocompatibility, non-toxicity, and chemical versatility.

In biomedical studies, it is used in drug delivery systems, wound healing, tissue engineering, and cancer therapy (Sahoo et al., 2025). Its mucoadhesive, antimicrobial, and hemostatic properties further enhance its biomedical value. Environmentally, CS is effective in removing heavy metals, dyes, and organic pollutants from wastewater through adsorption and chelation, making it a valuable tool in industrial effluent treatment (Liu et al., 2023).

CS broad applicability highlights its multifunctionality and sustainability. In this context, this chapter presents an overview of CS`s recent applications in the environmental and medical fields, highlighting opportunities and challenges.

Chitosan and Its Application in the Medical Field

Controlled Drug Release in Administration Oral, Parenteral and Nasal

Despite the widespread use of conventional drug delivery methods, such as oral administration or intravenous injection, these approaches often have significant limitations, including low bioavailability, rapid systemic clearance, and lack of target-site specificity. As a result, larger or more frequent doses are typically required to achieve therapeutic efficacy, increasing the risk of systemic side effects and compromising patient adherence. Furthermore, conventional formulations often fail to maintain optimal drug concentrations within the therapeutic window for extended periods, leading to fluctuating plasma levels and suboptimal clinical outcomes (Pereira, 2025).

In this context, the controlled release of drugs and bioactive molecules has become an alternative strategy in modern therapeutics, enabling the development of targeted delivery systems that increase treatment efficacy and minimize systemic toxicity (Pereira, 2025). These systems prolong therapeutic action at the site of interest, protecting healthy tissues and reducing adverse effects commonly associated with conventional delivery methods (Xiong, 2025). In this context, CS-based nanocarriers have attracted increasing attention due to their biocompatibility, biodegradability, and responsiveness to external stimuli, such as pH, temperature, or enzymatic activity (Vashitha & Khan, 2025).

Various routes of administration have been explored for the delivery of CS-based systems. Although oral and parenteral routes remain the most studied, alternative routes, such as nasal, ocular, and transdermal routes, have been increasingly investigated for their potential to bypass first-pass metabolism and enable noninvasive administration. For instance, Wu et al. (Wu et al., 2025) reported the development of pH-sensitive CS-alginate microcapsules for oral administration of *Ganoderma lucidum*, *Coicis Semen*, and *Lilium bulb* extracts. The material showed high encapsulation efficiency (91.82% ± 0.64) and exhibited sustained release with strong pH responsiveness, with 43.39% release at pH 1.2 over 30 min and 80.15% at pH 6.8. In another study, Shan et al. (Shan et al., 2024) reported the development of an artificial drug delivery system in polysaccharide microvesicles, uPA-CS/HS@RGD-ODE, to inhibit thrombotic recurrences. *In vitro* results suggest 61% specific targeting of activated platelets and responsiveness at pH 6.5. Furthermore, *vivo* results suggest that the carrier system extends the half-life of urokinase plasminogen activator (uPA) and allows its circulation time to be prolonged in the bloodstream. Finally, uPA-CS/HS@RGD-ODE exhibited superior targeting capabilities compared to free uPA.

Despite advances, oral administration of CS-based materials suffers from problems such as the formulations' vulnerability to digestive fluids. Changes in intestinal pH can affect CS protonation, resulting in reduced mucoadhesive properties. Furthermore, poor control of physiological variables and the lack of standardized mucus models create challenges for the implementation of these materials (Mehmood, 2025).

The parenteral route is a route of administration that bypasses the gastrointestinal tract by injecting medications directly into blood vessels or other tissues in the body (Mehmood, 2025). The injection and direct implantation of drug-loaded nanomaterials is an approach capable of generating positive results by promoting a marked increase in the effectiveness of treatments, reducing adverse effects, and allowing the sustained release of active components (Pouso, 2025; Li, 2024). The use of CS in the development of injectable nanoformulation allows for increased treatment efficacy, thanks to its functional groups that provide greater interaction between biological surfaces (Frank, 2020). These interactions prolong contact time with regions of interest and promote improved permeation of drugs with low solubility (Xiong, 2025).

The synthesis of a nanocarrier based on CS and iron nanospheres for the smart delivery of curcumin (Cur). The nanocarrier had an average size of 11 nm and a zeta potential of -30.7 mV. Furthermore, a loading efficiency of 64 ± 4% and satisfactory Cur release (> 80%) at pH = 7.4 were reported. The release kinetics followed a pseudo-second-order model, and cytotoxicity studies using VERO cells and *Artemia salina* demonstrated that the nanocarrier exhibits low toxicity. However, hemolysis at concentrations above

54 μmol L^{-1} was reported, suggesting a safe concentration for values below this. The authors report satisfactory interaction of the developed material with the protein albumin, suggesting potential transport through the circulatory system when administered intravenously (Fonseca et al., 2025). However, despite its promising nature, further studies involving animal models are needed to confirm its potential efficacy in living organisms.

In another study, the production of an implantable drug carrier based on ordered and graded porous ZrO_2 coated with CS has been reported. The fabricated system was investigated using ibuprofen (IBU) as a model drug, which achieved effective initial release control and prolonged effective release time. The authors report a maximum cumulative release rate of 93.7%, providing insights into the development of implantable drug systems. However, the study lacks further in-depth studies involving cell viability assays and animal model studies to confirm its biocompatibility (Wang et al., 2024).

Currently, researchers are exploring nasal route for systemic and central nervous system drug administration, due to its rapid and direct access to the brain (Alghareeb, 2025). Despite its convenience, there are challenges associated with using the nasal route that limit medication administration. These include the complexity of the nasal cavity, the presence and action of degrading enzymes, and the short contact time of the formulation with the absorbent nasal mucosa, which results in premature mucociliary clearance (Alghareeb, 2024).

CS stands out as a material with the potential to overcome the challenges associated with the development of nasal administration technologies. Its mucoadhesive properties extend the contact time of the drug with the nasal mucosa, reducing mucociliary clearance, and modulate the opening of intercellular tight junctions present in the epithelial barrier, increasing nasal drug absorption (Khan, 2017). For example, an *in situ* gel based on CS and niosomal methotrexate (MTX) was formulated for drug delivery to the brain. Niosomal MTX had a particle size of 130.5 nm, a zeta potential of -38.5 mV, and a trapping efficiency of 91.39%. Furthermore, compared with the free drug, the formulation showed sustained release and greater stability in the nasal cavity, resulting in a lower brain concentration (5.03 μg mL^{-1} vs. 1.96 μg mL^{-1}) and a lower plasma concentration (3.32 μg mL^{-1} vs. 4.87 μg mL^{-1}) of the drug (Ourani-Pourdashti et al., 2022).

Despite its potential for improvement, the use of CS has limitations in nasal administration. Among these, its reduced solubility in water and organic solvents, and its solubility at pH < 6 limit its application due to stability issues

(Shim & Yoo, 2020). In addition, its low encapsulation efficiency for hydrophilic drugs, combined with the low surface area for interaction in the nasal cavity, means problems in encapsulating the quantity required to achieve the therapeutic dose (Alghareeb, 2025).

Smart Biodressing

In wound treatment, dressings play a crucial role, with their effectiveness being crucial for rapid and efficient healing (Yang et al., 2023). CS-based dressings have attracted the interest of researchers due to their properties and the economic advantages of their use, compared to the more expensive raw materials used in conventional dressings (Seneviratne, 2024). They adapt to the needs and conditions of the different stages of the wound healing process, releasing therapeutic substances (Dong and Guo, 2021). These properties, combined with the growing demand for ecological alternatives, have driven the application of CS as a biodegradable polymer (Nguyen, 2023; Nuutila & Eriksson, 2021).

Smart CS-based biodressing was developed for the controlled release of green propolis (GP) for use in the treatment of dermal lesions. The material was developed using a design-of-experiment (DOE) and, in its optimal condition, presented a GP loading efficiency of 89.9 ± 1.1%. The prepared biodressing had a porosity of 88.7 ± 1.3%, water vapor permeability ranging from 74.3 to 98.3 × 10^{10} g ms^{-1} Pa^{-1}, and kinetics following a pseudo-second-order model. *In vitro* and *ex vivo* studies demonstrated efficient drug penetration into the skin and low toxicity to *Artemia salina*. The use of only naturally occurring materials with high biodegradability offers environmentally friendly and efficient alternatives for wound treatment (de Sousa Pereira et al., 2025).

In another study, a therapeutic hydrogel was developed using reuterin integrated into CS. The resulting material required no additional drug loading and exhibited tunable mechanical properties, pH responsiveness, tissue adhesion, and self-healing capabilities. *In vivo* studies demonstrated efficacy against bacteria, regulation of excessive inflammation, promotion of microvascular formation, acceleration of wound healing, and alleviation of scar hyperplasia (Gao et al., 2025).

Tissue Engineering

In the field of bone tissue regeneration, CS has emerged as a material with great potential for bioactive and osteoconductive support, which is easy to process into 3D structures with functionalizable and adjustable pore and surface sizes, with great potential for application in bone tissue regeneration (Ahmadian et al., 2020).

A hydrogel based on cross-linked CS and gelatin incorporated with zinc oxide nanoparticles (CG@m-ZnO) was also developed for application in bone tissue regeneration, exhibiting high compressive strength (973.31 ± 102.19 kPa) while maintaining microstructural integrity, exhibited excellent biocompatibility and high osteogenic differentiation of MC3T3-E1 cells, in addition to antibacterial activity (Hia et al., 2025). A photo-crosslinkable and injectable chitosan hydrogel grafted with methacrylic anhydride-functionalized kartogenin (CS-KGN-MA) was developed for cartilage regeneration, exhibiting controlled release of kartogenin in vitro and promoting chondrocyte-like matrix formation and satisfactory articular cartilage regeneration in rabbit tests (Ding et al. 2025).

3D-printed polycaprolactone/hardystonite composite scaffolds with CS-coated MnZn ferrite nanoparticles loaded with doxorubicin (DOX) were also fabricated for the treatment of localized tumors (Mohammadi et al. 2025). The authors evaluated the optimal scaffold composition based on its mechanical properties, biodegradability, and bioactivity. Furthermore, the feasibility of drug delivery by hyperthermia was assessed through radiofrequency exposure, indicating marked DOX release at 40 °C. These positive results demonstrate the potential of functional coating using CS in bone tissue regeneration and the treatment of localized tumors (Mohammadi et al., 2025).

In another study, Ghosh et al. (Ghosh et al., 2025) reported the development of CS-curcumin-citric acid nanocomposite scaffolds for the treatment of spinal cord injuries. The synthesized structure demonstrated controlled release of Cur for up to 72 h, low hemotoxicity, and *in vivo* studies indicated reduced spinal cord damage and possible neural tissue regeneration (Prabhakar & Pavani, 2025). In another study, a bioscaffold was developed based on guar gum (GG) and CS loaded with metoprolol succinate for tissue healing and regeneration. The synthesized material achieved controlled drug release over 24 hours and exhibited antibacterial activity. Cytotoxicity assays showed non-toxicity, and *in vivo* studies indicated cell adhesion to the scaffold, suggesting regeneration of damaged tissue (Ghosh et al., 2025).

Despite progress, these materials still have problems associated with low synergism between degradation rate and regeneration speed, as well as a lack of suitability for injuries to different tissues (Bai, 2022). Such problems need to be solved for a promising use of these materials in the future.

Chitosan and Its Application in the Environmental Field

In addition to its medical relevance, CS has attracted attention in environmental applications, particularly in water and wastewater treatment. Its natural abundance, biodegradability, non-toxicity and the presence of reactive amino and hydroxyl functional groups make it an attractive alternative to conventional synthetic polymers and inorganic coagulants (Kaczorowska & Bożejewicz, 2024). Its versatility enables its use in different remediation strategies, including the removal of heavy metals and dyes from wastewater (Tsauria, 2025), flocculation (Yang, 2016), filtration membranes (Liu, 2024), and even bioremediation and sustainable waste utilization (Chelu, 2023). This broad applicability highlights CS as a promising material in the design of green and cost-effective technologies for water purification and environmental protection. Below, we present an overview of these applications.

Wastewater Treatment: Removal of Heavy Metals and Dyes

The growing demand for high-quality water has driven the development of affordable and sustainable methods for the removal of contaminants from mining, industrial, and domestic effluents. Wastewater often contains heavy metals and synthetic dyes, which pose serious environmental and public health risks due to their toxicity, persistence, and bioaccumulation potential. Conventional treatment methods, such as chemical precipitation and ion exchange, while effective in some cases, present limitations including high costs, generation of secondary waste, and low efficiency in removing contaminants at trace concentrations (Ummalyma, 2024).

The application of CS and its derivatives in wastewater treatment becomes especially relevant considering that removal processes must comply with strict water quality limits established by international regulations. For metals such as cadmium (Cd), chromium (Cr), copper (Cu) and lead (Pb), even trace-level concentrations may exceed the maximum permissible values (OMS; EPA,

2024; Shofia, 2025). Table 1 summarizes the main regulatory limits for heavy metals in drinking water established by different normative authorities

Table 1. Regulatory limits for selected heavy metals in drinking water

Metal	WHO (mg L^{-1})	EPA (mg L^{-1})	EUD (mg L^{-1})	Brazil (mg L^{-1})
Pb^{2+}	0.01	0.015	0.01	0.01
Cd^{2+}	0.003	0.005	0.005	0.005
Cr^{2+}	0.05	0.1	0.05	0.05
Cu^{2+}	2.0	1.3	2.0	2.0

WHO: World Health Organization; EPA: Environmental Protection Agency; EU Directive: European Union Directive.

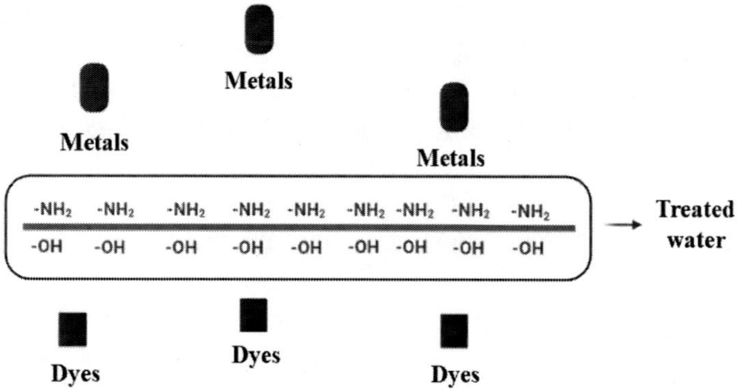

Figure 2. Proposed mechanism for heavy metal and dye removal by chitosan membrane.

Wastewater is also frequently contaminated with synthetic dyes, commonly originating from the textile, pharmaceutical, and food industries (Balachandran, 2025). These compounds pose significant environmental and toxicological risks due to their persistence, visibility in water bodies, mutagenic potential, and resistance to biodegradation (Balachandran, 2025). Dye removal can be achieved through several mechanisms, including adsorption, catalytic degradation, biodegradation, and advanced oxidation processes.

In this context, green and sustainable materials, including CS-based hydrogels, have been successfully applied for dye removal (Parida, 2025; Mohan, 2022). The chemical structure of CS, rich in amino and hydroxyl functional groups, favors the chelation of metal ions and electrostatic interactions with anionic dyes, enabling efficient removal even at low concentrations (Figure 2). Moreover, it is a biodegradable and renewable material that can be chemically modified to enhance its adsorption capacity, consolidating it as a promising resource in sustainable water remediation technologies (Yang, 2025).

A recent study reported the synthesis of a sustainable magnetic nanocomposite based on carbonized CS with Fe_3O_4 and SiO_2 for the removal of multiple heavy metals (Co^{2+}, Cu^{2+} and Ni^{2+}) from aqueous solutions. The material stood out for its ease of magnetic separation after the adsorption process and for exhibiting a high adsorption capacity, with a maximum value of approximately 2.908 mg g^{-1} (Ali and Ali 2024). In another study, the synthesis of CS modified with citric acid (Cit-CS) was investigated with the aim of enhancing surface active sites, stability in acidic systems, and the material's surface area. The modification resulted in a significant increase in heavy metal adsorption capacity, with maximum values of approximately 312.5 mg g^{-1} for Pb, 86.2 mg g^{-1} for Cu, and 8.2 mg g^{-1} for Fe. Moreover, the resin demonstrated good reusability, with desorption efficiencies ranging from 69 to 79% depending on the ion (Patel et al., 2025).

Similarly, in another study, CS and CS-based composites, such as CS/fly ash (CS/Fa), were evaluated for the removal of the textile dye congo red from aqueous solutions (Elhady et al., 2024). The experiments demonstrated that the pure chitosan achieved up to 98% removal efficiency, while the CS/Fa composite reached 94% under optimized conditions of pH, adsorbent dosage, dye concentration, and contact time. Adsorption data were best fitted by the Freundlich isotherm model, indicating a heterogeneous adsorption process, and kinetic studies suggested that the adsorption followed a pseudo-second-order mechanism, highlighting the potential of chitosan-based materials for sustainable wastewater treatment (Elhady et al., 2024).

Chitosan as a Flocculating and Filtration Membrane

Beyond its role as an adsorbent for the selective removal of heavy metals and dyes, CS also plays a significant role as a flocculant and filtration membrane. Due to its unique cationic nature and high density of functional groups (Figure 2), CS has been extensively studied as a flocculant, promoting the aggregation of suspended particles in wastewater and enhancing sedimentation and clarification efficiency (Nath et al., 2025; Kurczewska & Cegłowski, 2025). For instance, the combination of CS with graphene oxide (GO) has been reported for the removal of turbidity and multiple pollutants from surface water (Aboelfetoh et al., 2021). The GO/CS binary system exhibited high coagulation efficiency across a wide pH range, achieving turbidity removal greater than 98.3%, algae removal of 99.5%, and bacterial reduction at least 95%. In addition, the material demonstrated high removal capacity for organic dyes (direct blue 2 and methylene blue) and metal ions (Pb^{2+}), with adsorption kinetics fitting a pseudo-second-order model and equilibrium data following the Langmuir isotherm, reaching maximum adsorption capacities up to 667.8 mg g^{-1}. These results highlight that the combination of CS with nanomaterials like GO significantly extends its applicability and reinforces its potential as a sustainable alternative for water and wastewater treatment (Aboelfetoh et al., 2021).

In another study, Jia and Wang (Jia & Wang, 2023) investigated the combined application of polyaluminum chloride (PAC) and CS for treating low-temperature, turbid waters from the Pi River, Anhui, China. The CS/PAC coagulant at a volumetric ratio of 0.4:1 achieved turbidity removal of 86.32%, chromaticity reduction of 16.98%, and UV254 absorbance reduction of 32.79%, representing improvements of 17.9%, 3.77%, and 8.2%, respectively, compared to PAC alone. Optimal conditions involved rapid mixing at 300 rpm followed by flocculation at 150-100 rpm, with sedimentation time being the most critical factor for clarification (Jia & Wang, 2023).

In addition to flocculation, CS is also widely studied as a base material for the development of environmentally friendly filtration membranes. Traditional membranes used in water purification, while effective, are often made from non-biodegradable synthetic polymers, raising environmental concerns related to their production, use, and discard. In contrast, CS membranes are derived from renewable sources, aligning with the principles of sustainability and the circular bioeconomy (Azmi et al., 2025). These membranes can be used in microfiltration, ultrafiltration, or nanofiltration settings and can be adapted through chemical modification, polymer blending, or the incorporation of nanomaterials to enhance their performance. For example, the integration of CS with nanomaterials such as graphene oxide or metal-organic frameworks (MOFs) has been shown to improve membrane mechanical stability, antifouling properties, adsorption capacity, and selectivity toward specific contaminants (Liu, 2022; Li, 2022).

Functionally, CS-based membranes have demonstrated efficacy in removing heavy metals, dyes, suspended solids, turbidity, and emerging pollutants such as pharmaceutical residues. Their biodegradability, low toxicity, and energy-efficient production make them particularly suitable for integration into environmentally friendly water treatment systems, particularly in decentralized or resource-limited environments (Abdel-Raouf, 2023; Yang, 2025). Table 2 summarizes representative studies highlighting membrane composition, target analytes, removal efficiency, and environmental benefits.

Table 2. Studies reporting the development of chitosan-based membranes for water and wastewater treatment

Membrane material / Modification	Analyte	Removal efficiency (%)	Eco-friendly features	Ref.
CS/PVA cryogels	NO_3^-	62.5	Eco-friendly process	(Abou et al., 2025)
Chi-Py	As	96	High adsorption	(Venegas-García et al., 2023)
POMs	Ni^{2+}, Cu^{2+}, Pb^{2+}, CR and MB	>88% for heavy metal ions and >98% for organic dyes	Renewable and non-toxic.	(Fashapoyeh et al., 2025)
CS polymer modified with permanent cationic charges	As(V), Cr(VI), TC, CIP and AMX	98	Biodegradable and renewable biopolymer	(Esquivel et al., 2025)
Nanofibrous PVDF	Organic dyes and oily	< 90	Biodegradable and hydrophilic	(Yu et al., 2025)

CS: Chitosan; Chi-Py: Pyridinium-modified chitosan; CS-IMP: Copper-imprinted chitosan beads; POMs: Polysulfone–chitosan blend membrane incorporated with Keggin-type polyoxometalates; As(V): arsenate; Cr (VI): chromate; TC: tetracycline, CIP: ciprofloxacin; AMX: amoxicillin; PVDF: polyvinylidene fluoride; CR: congo red; MB: methylene blue.

These examples demonstrate the versatility of CS-based membranes and hybrid systems in addressing a wide range of contaminants while maintaining environmental sustainability. They highlight the potential of green filtration membranes as a promising strategy for advanced water and wastewater treatment, combining high efficiency with low environmental impact.

Conclusion

Chitosan has emerged as a highly versatile biopolymer with broad applications in both the medical and environmental fields. Its unique properties, combining biocompatibility, biodegradability, chemical functionality, and low toxicity, make it particularly suitable for the development of materials capable of addressing critical global challenges. In the medical field, CS has demonstrated significant potential in drug delivery, wound healing, and tissue engineering. Its ability to form a variety of structures, such as hydrogels, nanoparticles, membranes, and scaffolds, enables controlled drug release, enhanced bioavailability, and targeted therapeutic action. These characteristics establish chitosan as a fundamental platform in the development of next-generation biomedical technologies. In parallel, CS's high adsorption capacity and chemical adaptability have made it an effective material for a variety of environmental applications, including wastewater treatment, heavy metal remediation and dye removal. Additionally, the incorporation of CS into composite and hybrid materials has further enhanced their performance, selectivity, and reusability, providing sustainable and economically viable alternatives to conventional methods.

Disclaimer

None.

Acknowledgments

The authors are grateful to the Conselho Nacional de Desenvolvimento Científico e Tecnológico (CNPq, grants 405828/2022-5 and 408338/2024-5), and to the Fundação de Amparo à Pesquisa do Estado de Minas (FAPEMIG, grants APQ-01786-22, APQ-05429-23, RED-00144-22 and APQ-03853-25) for their financial support.

References

Abdel-Raouf MES, Farag RK, Farag AA, Keshawy M, Abdel-Aziz A, Hasan A. Chitosan-based architectures as an effective approach for the removal of some toxic species from aqueous media. *ACS Omega* (2023) 8:10086–10099.

Aboelfetoh EF, Aboubaraka AE, Ebeid EZM. Binary coagulation system (graphene oxide/chitosan) for polluted surface water treatment. *Journal of Environmental Management* (2021) 288:112481.

Abou Taleb MF, Aljowni MA, Parveen H, Mukhtar S. Green coagulation and flocculation: Scenedesmus algal extract-loaded chitosan/poly(vinyl alcohol) cryogel for effective water treatment. *International Journal of Biological Macromolecules* (2025) 290:138739.

Ahmadian E, Dizaj SM, Eftekhari A, Dalir E, Vahedi P, Hasanzadeh A, Samiei M. The potential applications of hyaluronic acid hydrogels in biomedicine. *Drug Research* (2020) 70:6–11.

Alghareeb S, Asare-Addo K, Conway BR, Adebisi AO. PLGA nanoparticles for nasal drug delivery. *Journal of Drug Delivery Science and Technology* (2024) 95:105564.

Alghareeb S, Ekenna I, Asare-Addo K, Conway BR, Adebisi AO. Chitosan nanoparticles for nasal drug delivery. *Journal of Drug Delivery Science and Technology* (2025) 105:106623.

Ali DA, Ali RG. Green synthesis of carbonized chitosan-Fe3O4-SiO2 nanocomposite for adsorption of heavy metals from aqueous solutions. *BMC Chemistry* (2024) 18:147.

Azmi LS, 'Ain Jabit N, Ismail S, Ku Ishak KEH, Abdullah TK. Membrane filtration technologies for sustainable industrial wastewater treatment: A review of heavy metal removal. *Desalination and Water Treatment* (2025) 323:101321.

Bai Q, Zheng C, Chen W, Sun N, Gao Q, Liu J, Hu F, Pimpi S, Yan X, Zhang Y, Lu T. Current challenges and future applications of antibacterial nanomaterials and chitosan hydrogel in burn wound healing. *Materials Advances* (2022) 3:6707–6727.

Balachandran B, Sabumon PC. A comprehensive review on biodegradation of azo dye mixtures, metabolite profiling with health implications and removal strategies. *Journal of Hazardous Materials Advances* (2025) 19:100834.

Chelu M, Musuc AM, Popa M, Calderon Moreno JM. Chitosan hydrogels for water purification applications. *Gels* (2023) 9:664.

de Sousa Pereira MV, Gonçalves Fonseca G, Pianeta Cervantes Y, Fontalvo Martinez JJ, Vieira EM, de Araujo Andrade T, da Silva RC, de Jesus JR. Development of a sustainable smart biodressing using chitosan, albumin, and green propolis for dermal lesion care: An in vitro study with multivariate analysis. *Journal of Polymers and the Environment* (2025) 33:2587–2601.

Ding Y, Zhu J, Guan X, Li C, Cai P, Lu C, Song L, Wu J. Photocrosslinkable hydrogel of kartogenin functionalized chitosan methacrylate preserves chondrogenesis. *International Journal of Biological Macromolecules* (2025) 321:146378.

Dong R, Guo B. Smart wound dressings for wound healing. *Nano Today* (2021) 41:101290.

Elhady S, Bassyouni M, Elshikhiby MZ, Saleh MY, Elzahar MH. Advanced chitosan-based composites for sustainable removal of Congo red from textile wastewater. *Discover Sustainability* (2024) 5:271.

Esquivel SE, Aranda FL, Vellojin JP, Pérez-Zurita D, Montoya-Rodríguez D, Villegas-Escobar N, Meléndrez M, Sánchez J, Palacio DA. Combining experimental and molecular computational approaches for enhanced removal of organic and inorganic pollutants via diafiltration using chitosan with permanent cationic charges. *International Journal of Biological Macromolecules* (2025) 320:145738.

Fashapoyeh MA, Shokrollahzadeh S, Isloor AM, Streb C. Efficient multi-pollutant removal: The role of Keggin polyoxometalates in polysulfone-chitosan blend membranes. *International Journal of Biological Macromolecules* (2025) 311:143738.

Fonseca GG, Vieira EM, Pereira MVS, Andrade TA, Santos AM, Gonçalves AP, Serafini MR, Santos AA, Figueiredo EC, Jesus JR. Smart pH-sensitive biopolymeric nanocarrier enables controlled release of curcumin through biomimetic biological barriers. *Journal of Polymer Science* (2025) 63:1628–1640.

Frank LA, Gazzi RP, Mello PA, Chaves P, Peña F, Beck RCR, Buffon A, Pohlmann AR, Guterres SS. Anti-HPV nanoemulsified-imiquimod: A new and potent formulation to treat cervical cancer. *AAPS PharmSciTech* (2020) 21(2):54.

Gao Y, Wang P, Huang S, Zheng Q, Wu S, Huang J, Chen Z, Gao B. Antibacterial reuterin as a multifunctional crosslinker for constructing chitosan-based hydrogels to promote infected wound healing. *International Journal of Biological Macromolecules* (2025) 316:144494.

Ghosh K, Dey I, Hazra A, Mondal S, Paul P, Bhowmick P, Bhowmick M. Fabrication and characterization of metoprolol succinate loaded chitosan-guar gum based bio scaffolds for infected wound healing and tissue regeneration: In vitro and in vivo assessment. *International Journal of Biological Macromolecules* (2025) 322:146714.

Hia EM, Park J, Suh IW, Park CH. Synergistic effects of modified zinc oxide nanoparticle in a hybrid chitosan-gelatin hydrogel for bone regeneration. *International Journal of Biological Macromolecules* (2025) 315:144490.

Jia R, Wang W. Performance of chitosan/PAC enhanced coagulation for low-temperature and low-turbidity Pi River source water. *Desalination and Water Treatment* (2023) 306:199–207.

Kaczorowska MA, Bożejewicz D. The application of chitosan-based adsorbents for the removal of hazardous pollutants from aqueous solutions—A review. *Sustainability* (2024) 16:2615.

Khan AR, Liu M, Khan MW, Zhai G. Progress in brain targeting drug delivery system by nasal route. *Journal of Controlled Release* (2017) 268:364–389.

Khoushab F, Yamabhai M. Chitin Research Revisited. *Marine Drugs* (2010) 8(7):1988–2012.

Kurczewska J, Cegłowski M. Tetracycline adsorbents based on alginate and chitosan – A promising approach to mitigate antibiotic resistance. *Journal of Water Process Engineering* (2025) 77:108518.

Li F, Yang Y, Yue H, Wang S, Zhang X, Wei W. Microneedle patch loaded with ferritin-nanocaged doxorubicin for locally targeted drug delivery and efficient skin cancer treatment. *Particuology* (2024) 88:282–289.

Li Q, Jiang S, Jia W, Wang F, Wang Z, Cao X, Shen X, Yao Z. Novel silver-modified carboxymethyl chitosan antibacterial membranes using environment-friendly polymers. *Chemosphere* (2022) 307:136059.

Liu X, Zou Z, Wang D, Du D, Li Z, Li W. Sustainable fabrication of chitosan membranes with optimized performance for ultrafiltration. *Separation and Purification Technology* (2023) 125276.

Liu Z, Wei W, Tremblay P-L, Zhang T. Electrostimulation of fibroblast proliferation by an electrospun poly(lactide-co-glycolide)/polydopamine/chitosan membrane in a humid environment. *Colloids and Surfaces B: Biointerfaces* (2022) 220:112902.

Mohammadi A, Mahmoodiyan Najafabadi F, Dini G, Ejeian F, Najafinezhad A, Ahmadi M, Michálek M. In vitro evaluation of 3D-printed polycaprolactone/hardystonite scaffolds with chitosan-coated DOX-loaded MnZn ferrite nanoparticles for bone tissue engineering and localized tumor treatment. *Journal of Drug Delivery Science and Technology* (2025) 114:107467.

Mohan K, Ganesan AR, Ezhilarasi PN, Kondamareddy KK, Rajan DK, Sathishkumar P, Rajarajeswaran J, Conterno L. Green and eco-friendly approaches for the extraction of chitin and chitosan: A review. *Carbohydrate Polymers* (2022) 287:119349.

Nath J, Dewan M, Ghosh A, Ray SS, Orasugh JT, Lahiri B, Chattopadhyay D, Adhikari A. Chitosan-based adsorbents for remediation of toxic dyes from wastewater: A review on adsorption mechanism, reusability, machine learning based modeling and future perspectives. *International Journal of Biological Macromolecules* (2025) 311:143388.

Nguyen HM, Ngoc Le TT, Nguyen AT, Thien Le HN, Pham TT. Biomedical materials for wound dressing: recent advances and applications. *RSC Advances* (2023) 13:5509–5528.

Nuutila K, Eriksson E. Moist wound healing with commonly available dressings. *Advances in Wound Care* (New Rochelle) (2021) 10:685–698.

Ourani-Pourdashti S, Mirzaei E, Heidari R, Ashrafi H, Azadi A. Preparation and evaluation of niosomal chitosan-based in situ gel formulation for direct nose-to-brain methotrexate delivery. *International Journal of Biological Macromolecules* (2022) 213:1115–1126.

Parida VK, Singh N, Priyadarshini M, Kumari P, Datta D, Tambi A. Insights into the synthetic dye contamination in textile wastewater: Impacts on aquatic ecosystems and human health, and eco-friendly remediation strategies for environmental sustainability. *Journal of Industrial & Engineering Chemistry* (2025) 150:247–264.

Patel PK, Pandey LM, Uppaluri RVS. Adsorption behavior and techno-economic analysis of citric acid-modified chitosan for heavy metal removal from wastewater. *Chemical Engineering Science* (2025) 315:121868.

Pereira MVS, Fonseca GG, Andrade TA, Gonçalves AP, Cruz KAML, Santos AA, Hespanhol MC, Despaigne AAR, Jesus JR. Amphiphilic nanocarrier based on poly(ε-caprolactone) for biomimetic membrane permeation and controlled lithium delivery: An in vitro study. *Journal of Polymer Science* (2025). https://doi.org/10.1002/pol.20250541.

Pouso MR, Melo BL, Gonçalves JJ, Louro RO, Mendonça AG, Correia IJ, Diogo DM, Hall ER. Injectable and implantable hydrogels for localized delivery of drugs and nanomaterials for cancer chemotherapy: A review. *International Journal of Pharmaceutics* (2025) 677:125640.

Prabhakar O, Pavani N. Bioengineering of chitosan/citric acid/curcumin nanofiber scaffolds for treatment of spinal cord injury. *International Journal of Biological Macromolecules* (2025) 321:146143.

Sahoo GP, Rajwar TK, Halder J, Mishra A, Mahanty R, Saha I, Satpathy B, Sahoo RK, Pradhan D, Rai VK, Dash P, Das C, Sarangi MK, Rout SK, Kar B, Ghosh G, Rath G. Natural bioactive compounds loaded chitosan nanocarriers: a promising strategy for breast cancer therapy. *Chinese Chemical Letters* (2025) 111797.

Seneviratne DM, Raphael B, Whiteside EJ, Windus LCE, Kauter K, Dearnaley JDW, Annamalai PK, Ward R, Song P, Burey P. A low-cost, antimicrobial aloe-alginate hydrogel film containing Australian First Nations remedy 'lemon myrtle oil' (Backhousia citriodora) – Potential for incorporation into wound dressings. *Heliyon* (2024) 10:e37516.

Shim S, Yoo HS. The application of mucoadhesive chitosan nanoparticles in nasal drug delivery. *Marine Drugs* (2020) 18(12):605.

Shofia SI, Vickram AS, Saravanan A, Deivayanai VC, Yaashikaa PR. Sustainable separation technologies for heavy metal removal from wastewater: An upgraded review of physicochemical methods and its advancements. *Sustainable Chemistry for the Environment* (2025) 10:100264.

Tsauria QD, Gareso PL, Tahir D. Systematic review of chitosan-based adsorbents for heavy metal and dye remediation. *Integrated Environmental Assessment and Management* (2025).

Ummalyma SB, Udayan A, Sreekumar N. Phycoremediation of emerging contaminants and heavy metals from industrial wastewater: A sustainable green approach for bioeconomy. *Bioresource Technology Reports* (2024) 27:101920.

United States Environmental Protection Agency (EPA). Lead and Copper Rule Improvements (LCRI). EPA; 2024.

Vashitha A, Khan SS. Recent advances in the development of chitosan-based nanocarriers for drug delivery application: Critical challenges, outlooks and promises in cancer therapy. *International Journal of Biological Macromolecules* (2025) 321:146184.

Venegas-García DJ, Steiger BGK, Wilson LD. A pyridinium-modified chitosan-based adsorbent for arsenic removal via a coagulation-like methodology. *RSC Sustainability* (2023) 1:1259–1269.

Wang L, He X, Hou X, Zhao Z. Preparation of CS coated graded ordered porous ZrO2 implantable drug carrier. *Ceramics International* (2024) 50:49963–49971.

World Health Organization (WHO). Guidelines for Drinking-water Quality. First Addendum to the Fourth Edition. Geneva: WHO Press; n.d.

Wu M, Zhang W, Song H, Hu Y, Li W. pH-responsive chitosan-alginate microcapsules as a delivery system for food-medicine homologous plant bioactives in lung cancer therapy. *Food Bioscience* (2025) 71:107364.

Xiong Y, Sun M, Yang Q, Zhang W, Song A, Tan Y, Mao J, Liu G, Xue P. Nanoparticle-based drug delivery systems to modulate tumor immune response for glioblastoma treatment. *Acta Biomaterialia* (2025) 194:38–57.

Yang M, Wang H, Li K, Chen Z, Seamirumi DT. A new soft tissue constructed with chitosan for wound dressings-incorporating nanoparticles for medical and nursing therapeutic efficacy. *Regenerative Therapy* (2023) 24:103–111.

Yang R, Li H, Huang M, Yang H, Li A. A review on chitosan-based flocculants and their applications in water treatment. *Water Research* (2016) 95:59–89.

Yang T, Gao H, Chen H, Xiao X, Zhao C, Gong H, Li X, Liu L, Liu Y. Insights and perspectives of chitosan-based hydrogels for the removal of heavy metals and dyes from wastewater. *International Journal of Biological Macromolecules* (2025) 292:139280.

Index

A

B

C

D

E

F

G

H

I

L

M

N

O

P